Super Radish and the Nut

Part One

Chapter 1

I am a radish, albeit a rather special radish of the species French Breakfast. Although I am much larger than the average for my species, I am very alike in appearance: oval, red on my top half and white below. My green leaves are a veritable panache, worthy of a very special radish. At the moment I am sitting by a rivulet dangling my feet (I did say feet) and prehensile tail in its cool waters. Glancing down narcissistically at my reflection, I am delighted by my shiny exterior, which has that healthy glow of a fresh radish.

You must be wondering how it is that a radish is

talking to you. Well, I shall explain. My life began in the laboratories of MonCul where the scientists were seeking to improve the shelf life of vegetables. I was Dr Frank's special project. He wanted a radish that would not wilt, shrivel up and look generally unappetising after days, weeks, months even on a supermarket shelf. To achieve this noble aim, my genome was modified to give me an impermeable skin, which would retain moisture indefinitely and thus prevent ageing. The process was a by-product of a program intended to stop human skin from ageing. The human experiment had the unfortunate side effect of freezing the human face in a set position, which made eating, drinking and osculation difficult because the skin became as hard and as tough as old boot leather.

My creator set about his task of producing the perfect radish. While he was at it, he decided to incorporate several novel features in my makeup. He was into nano-technology, which meant he could implant computers the size of molecules, millions of which could sit on the head of a pin. I was also given an array of nano-sensors. These would allow me to report on my progress, the state of the weather, surf the web, enquire about the health of my vegetable neighbours and transmit detailed reports, complete with mug-shots of any nasty anti-GM activists, who might try to interfere with the march of MonCul to world domination. My pride and joy is my complex eye, which is something like that found in insects, only better. It extends round the whole of my top half and permits me to see and photograph objects miles away. I can also focus on and enlarge details better than an electron microscope. How's that for a radish!

When I first became conscious of my surroundings, I was sitting on a soft bed of compost. I rustled my green panache and wiggled my prehensile tail to check their function. I was soon aware of a face peering down at me.

"Hello, Radish," he said. "I'm Doctor Frank. Welcome to MonCul and the world."

I didn't answer. I didn't like the look of his bald head, long, hooked nose and huge teeth, especially his huge teeth.

"What's the matter, Radish? The cat got your tongue?"

I maintained my silence. I didn't have a tongue; I was equipped with a speaker system. Dr Frank took me out of my soft bed and washed me in a bath of warm water. Over the next few days, I was prodded, squeezed and then dropped to test my dermal defences. I was subjected to blazing sun and freezing nitrogen baths and was not harmed.

"Age cannot wither you, nor custom stale your infinite variety," said Dr Frank, waxing poetical. My creator pronounced me perfect for long exposure on a supermarket shelf or an open-air stall. Then a thought crossed his mind: a long shelf life was one thing but edibility was a different kettle of fish. He decided I was inedible.

The composition of my skin became the object of Dr Frank's further research. My adamantine carapace suggested the possibility of use in the defence industries and large profits. Dr Frank attempted to cut off my leaves and tail and to pierce my skin with a scalpel and failed. I was dropped, used as a football and finally fired from a specially constructed gun with the aim of splattering me against a steel plate. I did not splatter. I passed through an inch of steel, the concrete wall of the laboratory and travelled a distance of 876 yards, bounced a further fifty-seven yards and came to rest in a dry ravine.

After swishing the dust from my complex eye with my prehensile tail, I took note of my surroundings, which consisted of red dust and rocks. I used my GPS facility to establish that I was in the Arizona Desert, outside the laboratory complex. That suited me fine because I did not wish to continue being manhandled by Dr Frank. I had my dignity as a radish to consider, and vegetables have feelings too. I remembered the bath I had been given on being born. There was no chance of bathing my eye in my present location: my energy needs were supplied by the solar panels in my glorious green tops; they needed washing too. I had to move to a more salubrious setting.

I tried to move. I rotated my prehensile tail in a rapid circular motion, rolled over several times and came up against a large rock. I needed transport. I scanned the desert and spotted a rattlesnake coming towards me. I couldn't see myself scaling the scaly snake in order to travel on the top deck; I would have to travel inside - snake, rattle and roll! The reptile paused and looked at me, trying to decide whether red and white indicated danger or merely slight indigestion. It made its mind up and struck, breaking its poison fangs on my diamond-hard carapace. What happened next shocked me. The snake writhed in agony and, in a few seconds, was dead. I had to find out why.

I settled down to surf the net to search for the MonCul laboratory records. Bringing the full power of my nano-computers to bear, it only took me a few seconds to access the secrets of my creation. One of my genes was implanted to protect me from insect pests. Judging by the untimely and painful end of the rattler, I was highly toxic. I had to think carefully about my next move. I spent the next few hours researching means of transport. Teaming up with an animal was out of the question; one lick from a friendly bear would lead to its demise. That, according to my ethical programming, was not allowed. Mechanical means were also beyond me. I had no limbs like a monkey or its cousins, the monkeys-without-tails, consequently I could not drive a car or a fly plane and I certainly could not ride a bike.

I then surfed the medical sites, particularly those on plant and animal physiology. I read up on stem cells, which are able to create body parts. I ruminated and rummaged internally and finally found a pocket of stem cells that didn't seem to serve any useful purpose except that of a reserve repair kit. Then I had to decide what would best suit a radish in terms of limbs. Monkeys had four, with which they could grasp objects: very useful if you lived in a tree. Monkeys-without-tails had two legs and unlike their cousins, walked upright only using their arms for things like eating and drinking. Walking upright on two legs was a hard skill to learn and some monkeys-without-tails fell over a lot. On one site, I read a slogan: 'Four legs good, two legs bad' so, to be on the safe side, I decided to go for four legs and four arms. That way it would be immaterial whether I walked forwards, backwards, sideways or diagonally; with my surround complex eye, no visible mouth, nose or ears, I would look the same from every angle.

I suffered considerable growing pains for the next week. I shepherded the stem cells to the sites on my body where I wanted arms and legs. I then put my computers to work urging on the cells. My first leg began as a small bump but soon stretched into the form of a very handsome limb. I chose the pattern from a site where the monkeys-without-tails were naked and mainly gathered in sets of four arms and four legs. Three more legs and four arms soon followed: I had my means of locomotion. I hear you asking. "So where do you intend to go? Where are you going to spend your vacation this summer?" I can only reply that I am but a few weeks old, so the world is my oyster to be opened and the pearl to be grasped. I spent several days exploring the world, and the universe even, on the web. I was surprised to see how little care the monkeys-without-tails were taking of their home. Judging them according to my ethical programming, they didn't care a jot about killing, maiming and starving millions of their fellows. They spent a lot of time and effort developing ways of killing their species. Strange creatures!

I eventually decided to go to the South Pacific Islands. I was attracted to them because they were sparsely populated, there was plenty of sun to supply my energy needs, and it rained every day, so keeping my eye and foliage clean would be effortless. I calculated that if I made my way to the Pacific Coast, I could hitch a lift on a piece of flotsam in the ocean and the currents would take me to my destination. I walked to the MonCul facility with great difficulty. The trouble with having four legs is co-ordinating their movements. I had placed them rather too close together and consequently kept tripping myself up and rolling along the ground in a tangle of legs and arms; most undignified. I solved the problem by retracting one pair and proceeding on the other in the manner of the monkeys-without-tails. Having reached the main gate, I waited for a vehicle to leave. I didn't have to wait long before a huge SUV roared to a stop. While the driver was being checked, I strolled to the vehicle and hoisted myself up into a comfortable position between the spare tyre and the back window. I was on my way.

The SUV travelled west to San Francisco to a house conveniently located on Long Beach. All I had to do was walk down to the ocean, select the best of the polystyrene lunchboxes that littered the shoreline, launch it and jump in. My Trans-Pacific voyage, I'm sorry to say, left a trail of dead fish. I tried warning them that I was toxic, but they would not listen and tried to take a bite out of me. Only the dolphins, intelligent creatures, paid any heed to my warnings and left me alone. My means of transport coalesced with a huge raft of debris, mainly consisting of plastic bottles and other packaging discarded by the monkeys-without-tails. It eventually beached on a small island and I was able to exercise my legs. I explored my new home, hiking to the top of the highest point, which was only a metre above sea level. I sat on a rock and consulted the

GPS. Having established the name of my island, I found the website, which described my new home, and got a shock: it was about to disappear.

I scanned 360 degrees and zoomed in on a group of natives on the beach. From their chatter in Pidgin English, I gathered that they were the last group to leave the island before it was reclaimed by the rising sea level. They were apparently heading for Nepal where they would be able to live out their lives safe from inundation and drowning. From my researches, I was aware that some scientists believed that the planet was warming up, causing ice caps to melt and sea levels to rise. The people at MonCul had dismissed this notion because, if they had accepted it, they might have had to stop using their two-miles-to-the-gallon SUVs. Just to be on the safe side, they were working on genetic modifications, which would allow monkeys-without-tails to develop gills, scales and fins and return to the ocean whence they had originally emerged.

I was deep in thought when I heard a voice. It came from a seashell. "What are you doing here?"

I focused on the mouth off the shell, zoomed in and identified the antennae of a Hermit Crab. "Show yourself," I said. "I feel stupid talking to a shell."

"No," said Crab, crabbily. "How do I know you are not from WorldCom Galactic Communications?"

"Because I am a GM radish escaping from the monkeys-without-tails," I replied.

"That's all right, then," said Crab, poking her

head out of her shell. "I'm escaping from those animals. I call them people, by the way. I was GMd at MonCul to become a relay station for WCGC's network."

"And how did they do that?" I asked, interested to learn how MonCul was abusing fauna as well as flora.

"My soft body was given a special protective coat, which allows me to live without a shell."

"Then why are you in that shell if you don't really need it?" I asked.

"I'm hiding, of course," said Crab even more crabbily. "As I was explaining, when you so rudely interrupted, I was given nano-implants."

"So was I!"

"There you go again, interrupting."

"Sorry."

"And new antennae. In short, I became a link in a global communications network until I found a vacant shell and defected."

"You do know that this island is about to disappear," I said.

"Of course, I do," said Crab glumly. "And I'm prepared to meet my end."

"Don't be so pessimistic," I said brightly. "I have no intention of throwing in the towel. I'm going to move on and you should come with me."

"Why should I?" moaned Crab. "MonCul gave me the Doomsday gene which means I can never have eggs."

"That's company policy," I countered. "They don't want people reproducing their GM products. If they allowed that, they would soon be out of business. We just have to face the fact that we are the end of the line, never to know the joy of our own offspring."

I noticed a tear trickling down Crab's pink

cheek. Unlike me, she had to breathe at intervals and needed organic food and a beach to survive. When the island disappeared, she would drown.

"Cheer up, Crab," I said. "You are not alone. I'll be your friend."

"Some friend," she grumbled. "A cross between a radish and an insect. The other crustaceans will laugh at me."

"You don't judge a sword by the scabbard," I said philosophically. "I am a perfectly formed radish with a few extra bits. I should warn you about one of my extra bits. Never touch me, because I am so toxic that I have been classified a WMD."

"Don't worry," said Crab sourly. "I have no intention of forming an intimate relationship with a radish."

"No need to be insulting," I protested. "You'll find me a great help when we embark on our mission."

"What mission?"

"Our mission to save the world," I shouted defiantly.

"What about the universe?"

"That too."

Chapter two

It took some time to persuade Crab to accompany me on my mission to save the world. She was of a naturally crabby disposition and called me a naïve optimist.

"How can you fight the power of the Multis and the World Organization who own the world, the Moon and now Mars?

"They are having trouble on Mars," I said. "A virus is cutting the invaders down to size. If a microscopic virus can beat the WO, so can we."

"Fat chance," said Crab. "But I'll join you on

one condition."

"Which is?"

"We have to take Coconut with us."

I was puzzled. I could see the nut Crab was referring to, lying under a tree. Coconut must have heard Crab, because he suddenly sprouted two legs, two arms and an antenna. A flap slid up to reveal a single Cyclops eye, and another to reveal a mouth filled with shining white teeth.

"Taraa!" said Coconut. "How's that for a metamorphosis?"

Coconut was evidently a well-educated nut, familiar with Classical literature.

"He was also part of WorldCom Galactic Communications," Crab explained. "He was supposed to monitor all wireless traffic and report back to Central Snoopers on any resistance to globalisation. He was going to report me as a defector, but I persuaded him not to."

"How did you do that?"

"I said I would get the Coconut Crabs to beat him up and, if necessary, chew him to pieces."

"Have you seen the claws and jaws on those beasts?" said Coconut, his voice trembling with fear. "They can slice through a coconut shell like a hot knife through butter."

"I wouldn't have let it happen, Old Buddy," laughed Crab. "You are far too useful as an early warning system." She turned to me. "Coconut knows over two hundred languages and is aware of everything that goes on in the world. He is the first to know when danger threatens."

"Well," I said, "danger now threatens in the shape of rising sea levels. We have to decide how to get away from here."

There was a long silence as we all concentrated

on the problem at hand. I made a mental list of our advantages, disadvantages and individual needs. Coconut, who insisted on being addressed forthwith as Coco (because he was such a clown) was able to shut down completely and roll into the ocean to float on the currents. He had been filleted of his coconut meat and as a result only needed to raise his antenna to capture energy from the microwaves, which swamped the ether. Crab was able to stay in the water for long periods, but had to find a beach to sustain herself by sifting sand. She was the weakest link in our chain. As a result of my added hardware, I was too heavy to float unaided and I had to use my green panache to capture the sun's energy. I couldn't do that at the bottom of the ocean. We needed a boat and the detritus on the beach was unsuitable as transport for our group.

Let me ask you a question. If you were on an island and your life depended on building a boat, could you do it? If you are flora: no. If you are fauna: possibly. The monkeys-without-tails can build anything, but all the other species labour under a handicap: they are hopeless at using tools. Crab was only able to manipulate sand into her jaws and hopeless at boat building. Coco had hands but wasn't adept at manual tasks: snooping was his forte. I had four strong arms and four strong legs. It was evidently down to me to do the heavy work. I consulted the articles about boat building on the web. Wood and harder materials were out of the question. Then I came across the reed boats of Lake Titicaca. We had reed-like materials on what was left of the island. I sketched a design in charcoal on a rock and we set to work. I used my four hands to plait grasses, which Coco gathered, into thick ropes. I used my computers to come up with a method for binding the ropes into a structure resembling the Titicaca model. One major difference was our decision to divide the boat into three sections. I would take the bow section so as to avoid brushing against Crab and inflicting instant death on her. We would put a layer of sand in the mid-section so that she would feel at home. Coco would take the stern section and maintain a permanent watch by monitoring the radio traffic.

For the launch, we had a little celebration. Coco tuned into a recording of the Berlin Philharmonic. To a rousing rendition of the *Ride of the Valkyries*, I climbed into the forward section, and Coco hoisted Crab into her sandy bed. He then pushed our nano-Ark into the ocean, leaping on board at the last moment as the surf lifted the bow. Our vessel rode proudly on the waves. All we needed was a stiff breeze from the land and we would be on our way east with the current. Using GPS, I plotted our progress, which was governed by the prevailing winds and currents. The consequence was that we moved generally east with periods when contrary winds drove us back. It gave us time to collect our thoughts.

Coco, the intelligence gatherer, was a fund of knowledge about the World Organisation. In general, the Multis controlled vast areas of the globe where they thought they could make good profits. They sent delegates to the WO, which was a holding company for the whole shebang and the real world government. Information on who headed the Multis and who they sent to the WO assemblies was not available. The WO only occasionally made a pronouncement. The invasion of Mars was a case in point. The Multis decided that it was time to invest huge resources in colonising the Red Planet, which had been probed and found to have vast mineral deposits. A company, MarsCorp, was given the trillion-dollar contract to carry out the colonisation. Although Mars was uninhabited, to justify the expense of the invasion, the WO PR department flooded the ether with announcements to the effect that the Martians held vast stockpiles of WMDs, which they could deploy within 45 minutes. The threat they posed was imminent and inevitable. The only logical answer, said the WO, was to carry out a pre-emptive invasion to prevent an attack by Martian terrorists. The first phase of the Mars campaign successfully established a dome, which tapped into the water beneath the surface and provided a liveable environment for the personnel. Then Mars struck back with a virus that decimated the invaders. MarsCorp headquarters in Washington received frantic calls for supplies of biological warfare suits, but by the time the supply rocket landed the garrison had been wiped out. The virus was the only life on Mars but that did not stop the WO from vowing revenge on the Martians. The lesson of the Mars War was cited whenever a pre-emptive war was subsequently decided on.

Crab listened to Coco's account of the state of the world and then waved a claw.

"That's all very sad," she said, "but what can we do against the might of the WO?"

"Knowledge is power," I said. "Between us we have a great deal of knowledge. It is only a matter of how we put that knowledge to use in fighting the WO."

"Where do we start?" said Coco.

"We start with MonCul," I said. "I have a feeling that it is much more powerful than is generally known."

"First we have to get to dry land," whispered Crab hoarsely. "I'm beginning to feel rather weak."

I was worried about Crab. She was the only one of us who got energy from organic materials. She had sifted through the sand at the bottom of her compartment several times; we had to find fresh sand. I used my sonar to measure the depth of the ocean. When I found a drowned island just beneath the boat, I suggested that Coco should lower Crab over the side to swim down and get a little nourishment. This worked well and Crab soon became her old crabby self, grumbling about how long we were taking to get to our destination.

We were being carried along by the Equatorial Counter Current, and my GPS told me we would soon reach the coast of Panama. When a palm-fringed beach came into view, Coco ran down the boat, hugged Crab and then me. Just in time, I stopped Crab from imitating him; I was probably still toxic. The surf cast us onto a beach of white sand. Coco lifted Crab out of the boat, which was pretty well waterlogged by now. Crab burrowed luxuriously into the damp shoreline and busily sifted sand for tasty morsels. I had always known that we would soon have to part company with her. She was not equipped for land travel and she had to have a beach on hand for her to survive.

Coco and I were strolling along the beach when he suddenly said, "We've got company!" He retracted his antenna and limbs and lay in the sand. I had no such camouflage, so I hid behind a rock. A girl of about ten arrived and looked down at Coco. She was about to pick him up when he opened his Cyclops eye and then revealed his flashing teeth. The girl shrieked and ran like the wind into the jungle.

"That was a cruel thing to do, Coco," I said. "She'll have nightmares."

"I suppose it was a bit drastic," he said, "but she was about to pick me up and I didn't fancy being bounced off a rock or chucked into the ocean." He chuckled, "I wonder what she is telling her mum."

We hurried back to Crab, and bade her a fond farewell, Coco with a hug and me with a wave.

"Keep in touch, Crab," I said.

"Of course," she replied. "Remember I'm a relay

station. I can communicate with you and Coco any time I wish."

I consulted my GPS and set a route for the MonCul facility in Arizona. There was a slight problem in that it wasn't marked on any map, but Coco was able to home in on their communications and use them as a beacon. We were near Panama City and could hear loud explosions. Coco listened in to the wireless traffic and found out that the WO was invading the country again.

"The WO appoints a president," said Coco. "He does as he is told for a while and then gets greedy and starts to rob the treasury. The WO then finds an excuse to depose him; it bombs the place, killing 2000 civilians in the process, arrests him and puts him in jail."

"I don't think we should travel overland," I said. "Those bombs are the neutron kind; they kill people and leave buildings intact. They would certainly damage our nano-electronics."

You're right," Coco said. "We'll have to get back in our boat and try to catch the North Equatorial Current. That should take us as far as Mexico."

When we got back to the boat, there was no sign of Crab so I got on the radio and contacted her.

"Don't worry," came her reply, "I've found a family of Hermit Crabs. They have a large spare shell. I'm going to stay with them."

"Crab has a family at last," I said to Coco. "I feel much better about leaving her."

The boat had been stranded by the tide and had dried out in the hot sun. We turned it over and emptied out Crab's sand bed. When we pushed it back into the ocean; it rode a little higher in the water. I resumed my place in the bow and Coco pushed the boat off, jumping into the stern at the last moment. The current caught us and we were soon making good headway north.

We came into Acapulco Bay as the sun was rising behind the Sierra Madre, bathing the ocean and the glistening white hotels on the seafront in a rosy glow. Coco and I leapt out of our boat. I had to use all four legs at first because we had been so long at sea and they lacked exercise. Coco strode confidently ahead up the beach and suddenly stopped. I caught up with him and joined him in gazing at an inert mass lying on the sand.

"It's a monkey-without-tail," said Coco.

"Is it dead?"

"I don't think so. I can hear it moaning."

The great mass levered itself into a sitting position. Coco and I stared at it in disbelief. There was no mistaking the bald head, hooked nose and tombstone teeth. It was Dr Frank, our creator.

"Where am I?" he groaned.

"You're on Acapulco beach," I said, "and judging by the lump on your bald head, you have been assaulted with a heavy instrument."

"Radish!" muttered Dr Frank. "Is that you?

"How many Radishes do you know?" I said sarcastically.

"But you've changed," he said. "I didn't give you four legs and four arms. And Coconut is with you."

"Hello, Doc," said Coco confidently. "Long time no see my innards."

"You both seem well," said Dr Frank. "Where have you been? What have you been doing?"

We enlightened the good doctor on our peripatetic vicissitudes. He listened intently.

"I must say," he said, when we had finished, "I

did pretty good work on you two. And I paid a heavy price for doing it."

Chapter Three

Dr Frank was not liked at MonCul. He was far too clever for his own good and didn't have the faintest idea of how to behave in a huge organisation. He passed colleagues and even superiors without greeting them. He wasn't deliberately rude, it was just that he was constantly preoccupied with his thoughts. He forgot to attend social gatherings and never made a fuss of the bosses' wives when he did. There were times when he forgot to leave his laboratory and worked into the night, thus convincing his colleagues that he was trying to show them up because they all left promptly at six.

His work on Coco was very successful. The nut was designed to monitor radio traffic, relay information to Central Snoopers and nothing more. Frank could not be blamed if Coco eventually defected. I was a different matter. Frank got carried away. Instead of just developing a vegetable with a long shelf life, he started to put in a few extras – a lot of extras. When I travelled through a steel plate and the laboratory wall, never to be seen again, Dr Frank found himself in very hot water. He had a final interview with his supervisor, Dr Gerbil Sneed, who was boiling with anger.

"I have your lab notes here, Dr Frank," he said. "What do you mean by giving a radish GPS, enough computing power to surf the net, sonar, hearing and speech facilities and goodness knows what else?"

"What can I say?" said Frank.

"When this interview is over, you can say, 'I resign'," said Dr Sneed viciously. "What have we got to show for all our investment?" He answered his own question: "Nothing."

"We have my lab notes," said Frank, pleadingly.

"And whose idea was it to test the radish to destruction?" said Sneed.

"Mine," said Frank.

"And the blasted vegetable was blasted out of the lab never to be seen again," said Sneed. "I'll expect your resignation on my desk by five today."

Coco and I listened to Dr Frank's sad tale quietly and felt rather guilty about the part we had played in his downfall.

"I'm sorry I defected," said Coco.

"I'm sorry I didn't return to your lab," I said.

"By the way, how did you get that bump on your

head?"

Frank ran his hand over his bald pate and winced when he felt the large swelling.

"It's a big as a coconut," he said.

"A walnut," said Coco.

"Whatever," said Frank. "I remember what happened now. I was taking an evening stroll along the promenade when I heard someone call for help from the beach. I saw it was a young lady who seemed to be in some distress. Forgetting the golden rule never to venture on the beach after dark, I ran down to render assistance. I was clobbered by four men who seem to have taken my wallet, my watch, my spectacles and my shoes."

"Why take your glasses?" I asked.

"They can get a dollar or two for the frames," said Frank. "It's the last time I play Good Samaritan."

Suddenly Coco retracted his arms and legs and collapsed onto the sand. Only the tip of his antenna was visible.

"Don't say a word," he said. "We are being monitored. I'll tell you when I get the all-clear."

I slumped against a cola can, one of the many that littered the beach, hoping that I would blend into its colour scheme. With the eyes on the top of my head, I followed a pigeon, which was circling overhead. I tuned into it and picked up a message.

"The target is on the beach. It looks as though he has been mugged. Returning to base. Over and out."

I watched the pigeon fly off with a whirring of wings. Funny pigeon, I thought. Coco's flaps flipped up revealing his Cyclops eye and tombstone teeth. I realised that Frank had copied his own teeth when creating Coco.

"What was the emergency?" Frank asked.

"A pigeon was observing you," I said. "I picked up a message to the effect that you had been mugged."

"That was a Stool-Pigeon," said Coco. "Central Snoopers deploy them to monitor anyone they don't trust. That was your personal pigeon, Dr Frank."

"I've noticed a particular pigeon ever since I checked into the hotel here," said Frank. "It's always the same one. I thought it liked my company. It perches outside my window and coos. I've tried feeding it, but it never eats my breadcrumbs."

"That's because it's a million dollar collection of nano-electronics," said Coco. "If you shot it, it would make a very funny pigeon pie."

"So MonCul are still interested in me," said Frank. "That's disgusting. Have we no right to privacy anymore?"

"You know as well as I do," said Coco, "that nobody's life is private. Once you have access to a Multi's secrets, you are on a watch list for the rest of your days."

A frown creased Frank's forehead.

"We'd better get back to my hotel," he said. "We have some thinking to do and plans to make. You two can't walk. People will notice if a radish and a coconut stroll into the hotel lobby. By the way, Radish, are you still toxic?"

"I don't seem to be," I said. "Several flies have landed on me recently and haven't dropped dead."

"Good," said Frank. "I gave your toxicity a limited life."

He picked up one of the many plastic bags that littered the beach. The one he placed us in was printed with the company logo and name of

BeefCorp, which had the world monopoly for cattle products. With the expansion of the world's population, using land for cattle ranching had become strictly limited. Meat had become a product affordable only by the very rich. Frank looked at the logo.

"I'm vegetarian," he said, as he carried us up onto the promenade, wincing when his bare feet found sharp stones. "I'm a vegetarian by choice, not just because meat is extremely expensive, but because I don't like the idea of killing animals for food."

We reached the promenade and heard someone shouting at Frank. Judging by the rude things he was saying, he was an employee of PoliceCorp.

"What are you up to?" he said. "Why are you barefoot? Do you know it's an offence not to wear shoes? It's a ShoeCorp regulation intended to keep the undeserving poor out of the urban centres. There's an on-the-spot fine of ten thousand pesos. Pay up. I take dollars and traveller's cheques but no personal cheques or credit cards."

"I can't pay up," said Frank. "I was mugged and robbed. If you will accompany me to my hotel, I will get you the money."

"Serves you right for going onto the beach at night," said the policeman. "It's people like you who lead our young people into crime. There's another ten thousand fine for that."

Frank knew better than to argue. A trip to the police station would mean a beating and an indeterminate stay in a foul, overcrowded communal cell. As we walked along, the policemen became quite pathetic.

"I have a terrible job," he moaned. "The pay is lousy and if we complain, we are transferred to one of the prison islands where there is a good chance of getting killed by the prisoners who have nothing to lose. What have you got in that bag?"

"Only an old coconut and a plastic radish," said Frank, trying to hide his alarm.

"Show me."

Frank reluctantly took out Coco.

"Why don't you put that in the trashcan?" said the policeman.

Frank dropped Coco back in the bag. "I'm going to use it to make a table lamp," he said.

I felt Coco shudder.

"Show me the plastic radish," said the policeman.

I was taken out and displayed.

"That's funny," said the policeman, who was enormously fat and very ugly. "A radish with four legs and four arms. I'll take it for my baby."

"I got it for my son," said Frank. "It's his birthday present. If I don't give him this for his collection of McDuff's toys, he will be very disappointed."

"How much did you pay for it?"

"It was free with a McDuff's Snappy Meal of tofu nuggets and fries."

"Well, I'm hungry," said the policeman who reverted to his ugly manner. "You can add the price of a triple beef-burger and large fries to my bill.

"I sighed with relief when I was returned to the safety of our plastic bag. The policeman must have heard me.

"Why did you make that noise," said the policeman. "Are you being disrespectful to a representative of the law? If you are, you are committing a serious crime."

"Not at all," said Frank. "The hot pavement is

hurting my feet."

"Serves you right for losing your shoes."

We arrived at Frank's hotel where cash was handed over. When we got into Frank's room, the pigeon was cooing on the windowsill. Frank went into the bathroom, taking us, still in the MeatCorp plastic bag, with him.

"This is the only place with no window," said Frank, running a hot bath, adding bubbles and slipping with a sigh into the healing warmth. "Now you two, you will have to stay in here. I know you won't be bored because you have access to all the entertainment and information you could want, built into your systems. You can start working on a strategy for me to survive into old age. You two have no worries since you are immortal."

Frank's last remark impressed me greatly. I

realised that it was true: I was immortal. The monkeys-without-tails killed themselves with junk food – the fools – and I had achieved a state that would allow me to study endlessly the whole of knowledge and become the greatest writer, poet, philosopher, scientist and radish ever to have lived.

Chapter Four

Back at MonCul, the CEO, Heinrich Graft had called a meeting of Heads of Department. He rose to speak, a grim expression on his face.

"I don't need to tell, you," he said, "that we are in a bind to the tune of two billion dollars. If this loss gets onto the balance sheet, we'll be bankrupt, heads will roll and some of us will do long jail time. After years of Patagonian Tooth-Fish and champagne, I for one don't fancy porridge and cold tea. You're the CFO, Dr

Trickie. What have you got to say for yourself?"

Richard Trickie, a sly-faced little man with eyes that each looked in a different direction, so that you were never sure whether he was addressing you or your neighbour, rose.

"There isn't a problem," he said in his wheedling tones. "I have already set up four shell companies and billed each of them for a billion, which means we have made a two billion profit rather than a two billion loss."

"That should keep the shareholders happy," said the CEO. "In the meantime, I suggest we all sell our stock and cash in while the share price is riding high. Not a word of this must get out of this room, of course. I shall arrange for Vanuatu passports for us all just in case we want to take a long vacation in the Western Pacific."

Angela, the eighteen-year-old girl who was the

Head and sole employee of the Corporate Governance and Ethics Department raised a timid hand and coughed: twenty-four faces turned on her.

"Yes, Angela?" barked the CEO. "Did you want to say something?"

"Well, yes," said Angela, in a sweet voice that echoed her name. "Is what Dr Trickie suggesting legal?"

Heinrich Graft became apoplectic.

"Are you suggesting that we would do something illegal? That is a very serious allegation. Can you imagine Dr Trickie doing anything illegal?"

"Well, yes," said Angela. "I have found out that he has already done four years in jail for fraud, embezzlement and insider trading, not to mention his strange habit of hanging round …."

It was Trickie's turn to become apoplectic and

leap up. "That's a lie!" he shouted. "I was framed and carried the can for a crooked CEO."

"Calm down, Trickie," said the CEO. "We all have our little secrets. Don't worry. You're doing a great job and this will be reflected in your end-of-year bonus. Sit down, Angela. You have had your answer. I want to hear from our Contracts Manager."

Rupert Bach-Hand, the Contracts Manager was a jolly, florid-faced, overweight bon vivant, who had successfully negotiated MonCul's most lucrative contracts. His expense account was greater than the GDP of Liberia and he had ways of persuading the decision-makers of the WO to put multi-billion dollar contracts MonCul's way. He avoided the effort of standing up on medical grounds.

"Good news!" he said. "You can take those grim

looks off your faces. An hour ago, I got a message from my inside contact at the WO that we have been given the contract to develop Adamantine, the material of the future."

The meeting applauded. Heinrich Graft smiled. Angela tried to make herself smaller. Rupert beamed.

"That means," he said, "one hundred and eighty billion dollars up front for Research and Development."

Heinrich Graft could not contain his joy. He leapt out of his chair, ran to Rupert and shook his hand. "Well done, Rupert." He turned to Dr Trickie. "Now, what about the accounts, Trickie? When that cheque thuds into the balance sheet, we'll be laughing."

"I haven't finished," said Rupert. "As you may know, Adamantine is a material, which will have many uses, such as battlefield whole-body protection; it will provide impenetrable armour for our APCs, tanks, aircraft and warships; missiles, shells and bullets with adamantine coating will also penetrate the hardest defences."

Mary Stocking raised her hand to speak. She was an attractive woman in her mid-twenties with a PhD in materials science.

"Yes, Mary," beamed Heinrich. "You have a question?"

"I do indeed," said Mary. "I was hired a month ago to work on a new material. Is that what is being referred to?"

"Yes," said Heinrich. "That's the stuff: Adamantine. I gave you the late Dr Frank's notes. Reproducing his manufacturing process should not be a problem."

"But it is a problem," said Mary. "I have repeated his methodology a hundred times and the nearest I have got to a hard material resembles treacle."

"That will keep you sweet!" said Heinrich, laughing. The meeting roared with laughter. "Mary, Mary, don't be contrary." (More hilarity) "You have to be positive."

"I am positive," said Mary. "I'm positive that a conjunction of unrepeatable circumstances led to the creation of the indestructible radish. Either that or there are some gaps in Dr Frank's lab notes."

"You are new here, Ms Stocking," said Heinrich indulgently. "You need to learn that we operate as a team and must always maintain a consensus. To quote a president long dead, 'If you are not with us, you are against us. If you are not against the terrorists, you are a terrorist.' Look at the Mars War. We failed, because of the doubters. We were beaten by them and the terrorists using WMDs."

"It was a virus," said Mary.

"A biological WMD," said Heinrich. "The Martians won but we will never forget. We shall return and win."

The meeting erupted in patriotic applause and an impromptu rendition of 'God Bless MonCul' was sung.

"There aren't any Martians," said Mary, drawing a gasp from the meeting.

"Our dead heroes may not have seen them," said Graft, his face like thunder, "but they are there and we will find them. They can skulk in their caves, but we will bomb them out. By the way, Stocking is funny family name. Is it real?"

"As real as Smith, Carpenter or Graft," said Mary coolly. "By the way, you're not related to Herman von Graft, the mastermind behind the African genocide? Of course not. And while we are on names, MonCul is a joke in French, where it means something rather rude. You should change it."

Heinrich was livid. "The name is based on Montana and Culver Town where our great company was started in a garage by two young biologists." Satisfied that he had squashed Mary Stocking, he went back to his seat. "Any other business?"

"Yes," said Mary. "Where can I find Dr Frank?"

"The last I heard of that idiot, he was down and out in Acapulco, lying unconscious on a beach," said the CEO. "I believe he is in the last stages of cirrhosis and due for a massive, fatal, portal haemorrhage at any moment."

"That's strange," said Mary. "I met him at a conference last year and he was very well. He's a vegetarian and he doesn't touch alcohol."

"I've had enough of your impertinence, Ms Stocking," barked the CEO. "People change!

This meeting is closed." He went over to Rupert Bach-Hand and muttered, "Come with me. We need to talk."

Back in his penthouse office, Heinrich poured refreshment for himself and Rupert.

"Tell me, Rupert," he said, "what did it cost us to get that R and D contract?"

"Four directorships at two million a year in the new shell companies," he replied. "Cheap at half the price."

"Yes," said Heinrich. "I couldn't admit it in the meeting, but Contrary Mary had a point. We don't have the product and I doubt whether we will ever develop it. It's too much like the Philosopher's Stone."

"Don't worry," said Rupert. "We have locked our contract into a deal with AirCorp to build a bomber that is invisible to radar and can't be shot down because it is coated with Adamantine."

"Will they build it?"

"No way," said Rupert. "And the lead time is twenty years, by which time we will be long gone to luxurious retirement."

"Good," said Heinrich. "Do you think we should bring Dr Frank back to silence that Stocking woman?"

"Bad idea," said Rupert. "But you have to get the lady out a.s.a.p. You were a little too specific about Frank's impending demise and she is suspicious. What are your plans for Frank?"

"SolutionsCorp was going to eliminate him but we will have to call it off," said Heinrich. "It will be the Chairman's decision, of course."

"Have you seen him recently?"

"The Chairman? No. I have never seen him," said Heinrich, "but the decisions come down from Culver every day."

"Don't you think it odd that nobody has ever seen him?"

"Not at all," said Heinrich. "He has a completely logical mind and always comes up with solutions free of any emotional attachments. He's brilliant. He doesn't need meetings like the one we have just witnessed."

Chapter Five

Coco and I sat on the edge of the bathtub and dangled our feet in the warm water, while Dr Frank soaked in bubbles and dreamed.

"Why is it, Frank, that you spend so much time bathing?" asked Coco.

"Archimedes made his finest discovery in the bath," said Frank. "I find I can relax and think clearly immersed in warm suds."

"Are you going to suddenly leap out shouting 'Eureka'?" I asked.

"Probably not," said Frank. "I don't want to slip on suds and break something. Anything of interest on the grapevine, Coco?"

"According to an e-mail I have just intercepted, Mary Stocking, your successor at MonCul, is about to be given the push after only a month."

"Why?"

"Gross insubordination, unsatisfactory work and dissolute behaviour," said Coco.

"Mary is a polite, intelligent, honest woman who doesn't know how to misbehave. Dissolute behaviour, indeed. That's poppycock. She probably spoke out of turn at a meeting," said Frank. "Was she getting anywhere with her research?"

"No progress to report," said Coco.

"I've made a decision," said Frank suddenly.

"We're leaving this hotel and my personal pigeon isn't going to follow us. I've bought something for this occasion."

He emerged from the bath, dried, and dressed, and went into the main room. He upended a plastic bag and spilled out a tube of adhesive, a lead box and a hammer. He went to the window and spotted his personal pigeon circling above. He opened the window, spread the adhesive on the sill and waited. True to form, the pigeon landed on the windowsill and cooed. When it tried to walk, it couldn't move. It flapped its wings, and they, in turn, were trapped by the adhesive. Frank popped the lead box over the pigeon and hammered it flat. The crunch of electronic parts and motors was audible.

"Home at last, pigeon," said Frank. "I only hope you didn't get off a message before I got the lead box over you."

When Frank checked out of the hotel, with Coco and me in his bag, there was an emergency at Central Snoopers. The Pigeon Patrol Duty Officer immediately contacted maintenance to ask them to find out why SP007 had ceased transmitting. Coco picked up the radio messages and established that SP007 had not been able to get off a distress call before being flattened.

"That's a relief," said Frank. "We need two hours to get to my redoubt in the Sierra Madre. Let's hope Central Snoopers fail to track me before we are home."

We went into a disused building. When we emerged, Frank was dressed as a female Mayan Indian, Coco was swaddled like a baby in his poncho and I was the pacifier in baby's mouth.

"Don't chew, suck!" I said to Coco, who was

playing the part of a baby like a method actor.

"Sorry," said Coco.

"Dr Frank," I said, "did you know you were being watched by Central Snoopers?"

"No," said Frank, "but they always put people like me on a watch list. We are now going to take a bus into the mountains. You will like my hacienda."

"So you don't live in hotels," I said.

"No," said Frank. "I was only in Acapulco to get some cash from my account."

Frank bought a first class ticket and we boarded a bus crowded with farmers heading back to their mountain villages. They were mostly men who kept ogling Frank with toothless grins. I must say Frank, in a black wig and colourful Mayan attire, looked much better than his normal self. We jolted along with several stops to let the second class passengers get out to push the bus up particularly steep hills. I realised then why there were two classes on the ramshackle bus. We reached the hacienda as the sun was setting. A Mayan woman, who had evidently heard the bus arriving, how could she not, rushed out to carry Frank's bag into the kitchen.

"Welcome back, Senor Martinez," she said. "I hope you had a good trip."

So Frank was Senor Martinez in Mexico. He took off his wig and bent to display the angry bruise on the top of his bald skull.

"I had a problem Teresa," he said."

"I'll make a tortilla poultice for that," she said. "You should always wear your Mayan dress away from home; then the bandits won't touch you."

"I know," said Frank or Senor Martinez. "But I was in a hotel and had to visit the bank. Let me introduce my guests." He opened his bag and took Coco and me out and placed us on the kitchen table. "Don't be afraid," he said, "These are like toys." (Toys? Frank could be thoughtless on occasion.) "Coco, Radish, wake up!"

I stood up and did a little four-step dance. Teresa laughed. Coco extended his arms, legs and antenna, and lifted his flaps to reveal his Cyclops eye and tombstone teeth. Teresa screamed.

"Hello, Teresa." Coco and I said in unison.

Teresa crossed herself, sat at the table and stared at us. "What wonderful toys. They must be made in China. But I'm forgetting my duties. You are hungry, Senor. I have prepared guacamole, tortillas, vegetables fresh from the garden and salsa."

"Lovely," said Frank. "I'm starving."

As he was eating, Frank explained that he grew all his own vegetables organically. "When I was a student," he said, "I did research on plant physiology. I discovered that plants, particularly flowers, reacted to threats and potential danger. I was able to measure this by attaching electrodes to them and observing the variation in electrical activity in them when approached with a knife or scissors. I was already a vegetarian, and I began to wonder whether I was inflicting pain on salads by eating them raw. I thought about only eating vegetables that had died of natural causes, but yellowing cabbage leaves and wizened carrots taste rotten."

"Well," I said, "speaking as a radish, I can tell you that vegetables don't particularly relish being peeled, chopped, boiled, fried or stewed and eaten. I am much preoccupied with philosophical questions these days. In the past, theologians pondered whether animals, as sentient beings, had a place in Heaven. Would Spot be joined by his master, or vice-versa, in that great pleasure garden in the sky? I have asked myself, where do vegetables stand in the queue for Paradise? If you cut a beetroot, does it not bleed? Doesn't a parsnip have the same value as a rabbit or a gerbil?"

"Well," said Coco, "I don't know how big that place in the sky is, but, if you are right, it will be knee deep in onions and goodness knows what else."

"Since you have been deprived of your meat, Coco," I said, "I don't expect you to have the same fine feelings as a lettuce with a big heart."

I was just about to indulge in a fine sarcastic sally, when the door opened and in hobbled a chicken with four legs holding a white stick in one claw.

"This is Dora," said Frank. "She's come to welcome me home. Hello, Dora"

Dora clucked her welcome.

"She's the saddest example of science gone mad I have ever seen," continued Frank. "The idea was to produce a chicken with four drumsticks so that each member of the average family of two adults and one point five children could have a drumstick each. The saddest thing of all was that Dora was created blind to stop her chasing round her cage and losing weight. I rescued her from a McDuff experimental farm."

"I saw a tear trickle from the corner of Coco's eye. He was a sensitive nut after all, in spite of being filleted. It was with a catch in our voices that we bade each other goodnight. Dora went to her basket in the kitchen, Frank went to bed, Teresa did the washing up, and Coco and I switched off for the night to let our systems cool down.

The next morning, we were awakened by Dora crowing. That was another sad feature of the Dora experiment; she suffered from a split personality and didn't know whether she was a cock or a hen. Frank had chilli-fried eggs with chocolate mole for breakfast.

"Caught you out," I cried. "Vegetarians don't eat eggs!"

"No, you haven't," said Frank, the casuist. "These eggs are unfertilised and therefore could never have become chickens. Dora lays them."

He showed us round his vegetable and fruit farm. In the GM corner, he pointed to examples of designer vegetables: square tomatoes, onions and cucumbers, ideal for sandwich making. There was a cabbage crossed with a carrot which, when shredded, made excellent coleslaw.

He had potatoes which produced bunches of skinless chips, ready for washing and instant frying. For children who did not eat spinach, there was a variety, which tasted like bananas. There were seedless apples the size of strawberries and sweet, seedless, hairless gooseberries the size of apples. All the fruit and vegetables were seedless because of the Doomsday gene they contained.

We went into Frank's potting shed to avoid being overheard by Teresa.

"She's a good housekeeper," said Frank, "but she does like to chatter to her friends."

"Dr Frank," said Coco, "Teresa knows you as Senor Martinez. I think we should use your given name. What is it?"

"Francis," said Frank. 'My full name is Francis Frank. You can call me Frank."

"So you're Frank Frank," I said. "Your parents didn't have much imagination."

"I've no idea," said Frank. "I never knew them. I was found in an abandoned self-propelled supermarket cart near the Golden Gate Bridge in San Francisco. The orphanage named me."

"Have you never tried to find your parents," said Coco. "It might be possible to trace them on the National DNA Database."

"I know that," said Frank, "but I reckoned they did not want to keep me, so I wasn't going to embarrass them by turning up on their doorstep."

"If you were found in a self-propelled supermarket trolley," I said, "don't you think that you could have moved it yourself. They can travel a hundred miles between battery charges."

"I don't think so," said Frank. "I was only six months old at the time."

We dropped the subject of Frank's sad separation from his parents and turned to more urgent matters.

"Frank," said Coco, "I think you should get in touch with Mary Stocking and warn her that her life is in danger. MonCul will never allow her to blab about the failure to reconstitute Adamantine."

"I've already come to that conclusion," said Frank. "But how can I contact her without revealing my location and leading a SolutionsCorp hit-man to me? I was tracked to the hotel in Acapulco because I was forgetful and foolishly registered in my real name. That Stool Pigeon was on me immediately."

"I have the same problem," said Coco, "I can monitor the world's wireless traffic, but anything I send will be monitored too. What about you Radish?"

"Same here," I said, "but I have an idea. I could get Mary to join a chat room and develop some kind of code so that we can arrange a meeting."

"It's not feasible," said Frank, "DecryptCorp can crack any code. We can't use electronic means."

"Radish," said Coco, "how did we get across the Pacific to Panama and then up to Mexico?"

"In our reed boat," I said, "but such a craft won't carry Frank."

"Of course not," he said. "Frank can get a boat."

"That's an idea," said Frank. "Teresa has a brother in Acapulco who owns a fishing boat. Perhaps I could charter it. I'll talk to her."

Chapter Six

Mary Stocking was relaxing at her parent's home in San Francisco, when the doorbell rang. It was a Delivery Hound from ExpressCorp with a letter. She took the letter from the hound's pouch located under its jaws, signed the receipt and placed it in the pouch. If anyone other than the designated recipient were to try to take a letter, they would have their fingers bitten off. Mary patted the hound on the head but it didn't leave.

"Oh," she said, "I forgot your tip. So sorry."

She went to the kitchen and returned with a biscuit, which she placed in the pouch labelled 'Gratuities'. The huge beast trotted off to make its next delivery. Mary didn't like the Delivery Hounds, operated by implanted chips, which directed them to the customers' doors. They had been developed to replace the human couriers who had gone on strike once too often for ExpressCorp's liking. Now the couriers were all out of a job and had been transferred to Under Class Areas where they were isolated from the community at large. Mary was a little afraid of the Delivery Hounds, some of which were getting aggressive and attacking people who refused to give them a gratuity. As for the redundant couriers, they were trucked out of their ghettos daily, the women to work as domestic helpers, the men to do menial jobs like street cleaning, refuse collection and road construction.

Mary went into the kitchen and examined the letter. She tested it with her portable BiChem Scanner to check that it wasn't a biological or chemical WID (Weapon of Individual Destruction). Because of the acronym, these were commonly referred to as the 'widow makers'. It was clean. Mary used a paper knife to open it. If the envelope was impregnated with high explosive, tearing it open would detonate it. It didn't explode. Mary had been taking such precautions since being thrown out of MonCul. She assumed she had been placed on the Subversives Watch-list, which made her fair game for any madman from SolutionsCorp who needed a bonus for a fresh hit. The letter was written in ink by a shaky hand, probably that of an old person. It read:

My Dear Grand Niece,

It is so long since I have heard from you; frankly, I am beginning to think that you have forgotten all about me. Some time ago, I saw you on the TV at that conference you attended and I thought you looked lovely. You were having a frank exchange with a person who was worried about how science was perverting our flora and fauna.

I shall be coming to San Francisco in the afternoon of the 13th of next month and I would love to talk to you again while we stroll in the *park. You know that I hate hotels. Could you put me up (and put up with me) for a couple of days? I know the weather will be cold, so I'll wear my Mexican poncho that your grand uncle brought back from Mexico.*

Your loving Grand Aunt, Frances

Mary didn't have a grand aunt of any name, and the words frankly, frank and Frances jumped out of the page. She had had a very frank exchange with Dr Frank at the last GM conference about the way science was going. He had struck her as too idealistic and rather eccentric. In spite of that she had liked him for his commitment to ethical science. She knew she couldn't let him come to her parents' home. There was a pigeon that watched all her movements and all her contacts. She racked her brain trying to think of a solution. The return address on the letter was in

Swahili and translated as Third Mosque on the Left, Timbuktu, so that was no help. She would have to take the risk and let Dr Frank arrive and hope that he would not be observed. It was very worrying. She was just about to tear up the letter when the ink began to fade and then disappeared.

On the 13th, Mary left the house in the afternoon and walked in the park. She had guessed the reference to a Mexican poncho was important, so when an old woman wearing a poncho and a voluminous ethnic headdress approached her, she followed her into the underpass, which led out of the park. The woman stopped and held out her hand as if asking for alms. Mary opened her purse and handed over a small bill. She felt the woman thrust a tightly folded piece of paper into her hand and then she walked away. Mary waited until she was back home in her bathroom before she read the note. It directed her to the Hispanic quarter of the city and told her to go to a particular outlet of Megabucks Coffee at five the next day and loudly ask for a pot of tea. As with the letter, the ink on the note faded and disappeared. She screwed it up and flushed it down the toilet.

Megabucks was crowded, and she had to wait in line to get to the counter.

"A pot of tea, please," she said loudly. "Preferably Orange Pekoe tea."

The youth serving her laughed. "Tea, lady. We don't got no tea here. This here is Megabucks. We don't serve no tea."

"In that case," said Mary, "if I can't have tea, I'll take an espresso. I was so looking forward to a nice cup of tea."

She took her cup and found an empty table at the back of the room. Before long a young woman carrying a large shopping bag approached her table and asked if the other seat was taken. Mary shook her head and the young woman sat down. As she sipped her coffee, she said, "When I go to the restroom, wait two minutes and then follow me."

The young woman left. Two minutes later, Mary went to the rest room and saw the young woman at the washbasins. She stood alongside her.

"Take the bag," whispered the woman. "Change in a stall and take your things with you. There's a Great Wall minibus parked outside. Get in. It will take you to your destination." With that she walked out.

When Mary emerged from the restroom, she was wearing a burka, which covered her from head to foot. She left the coffee shop and got into the waiting bus. The driver said nothing. It was left to Mary to ask where they were going.

"The Bay Under Class Area," said he driver.

"I thought vehicles were prohibited from entering there," said Mary.

"Don't worry, this is an official bus and I'm on an approved pickup," said the driver. "Three women at a house in Glendale. I'll tell them you are new in the ghetto. You'd better change back into your own clothes."

He said no more until he stopped at a fortified compound to pick up the three women who were watched over by a security guard. They got into the bus and greeted the driver.

"Who's the new girl, Bernie?" asked one.

"Mary's just joined our exclusive community," said Bernie.

"What was the charge, Mary?" the woman

continued.

"Leading a dissolute life, incompetence and insubordination," said Mary.

"You told the boss to drop dead, didn't you?"

"Something like that."

The women, who were hired out as cleaners, were silent, all thinking about the petty crimes which had led them to the Bay Under Class Area.

"Here we are," said Bernie. "Home, sweet home."

Bernie held out four ID cards, one of which was for Mary, now Maria Martinez, to be checked, and was waved through the gate.

Over the half-century of their existence, the UCAs had become practically autonomous and were no-go areas for PoliceCorp. Once you were in the settlement you had little chance of living outside. In the early days, if parents wanted their children to live outside, they had to give them up as infants for adoption by rich couples who no longer reproduced on the grounds that it was better left to the poor. That practice soon stopped. The UCA was, in fact, a ghetto, where all the work was done by the inmates. They ran community services like utilities, education and medical care. They built and maintained the fabric and infrastructure. They only went outside if there was a need for workers to do menial jobs and they had to return every evening. Some ran away but were soon tracked down because they had an implanted chip.

The sun was setting, but Mary could see that the settlement was not a bit like she had expected. The streets were swept, the houses were neat and well maintained and there were trees and green sitting-out areas everywhere. There were schools, a hospital and municipal buildings.

Bernie could see that Mary was bemused by what she was seeing.

"We run our own affairs," he said. "We levy our own taxes and balance our own budget. The Federal Government hates us and says we are a bunch of communists. The main thing is that it stays out of our hair. We don't pay any federal or city taxes because, in theory, we are prisoners. Any money we raise in taxes is spent here and nothing goes to fight wars like the Mars debacle. And you won't see any pigeons here. We shoot them down."

The minibus stopped and Mary was led into a neat villa surrounded by trees, lawns and flowers. Dr Frank was waiting for her in the living room.

Chapter Seven

Herman Spandau had been an astronaut who had perished on a mission to Venus. Before his demise, he had named his infant sons Wilbur and Orville in honour of the first great aviators. Their education had been provided for from a generous pension awarded to their widowed mother and they emerged from Harvard with PhDs in biology. As was the case with most gifted young men, they went into business on their own account, in their hometown, Culver, Montana. They raised a million dollars from a venture capital fund and started work in their garage developing eco-friendly insecticides. Their research was very fruitful and they soon had a large royalty income from their products. Flying was the brothers' passion, so they decided to go into the crop-spraying business using their own products. They were soon employing a dozen pilots to operate their fleet of VTOL state-of-the-art crop spraying machines. They became rich.

As dedicated scientists, the Spandau twins were incapable of sitting back and enjoying their wealth: they had to pursue new research. They went into nano-technology and genetic research and eventually combined the two disciplines. They would have understood why Dr Frank got carried away when he was working on me. They developed a potato, which was gradually transformed into an organic mega-computer. It needed no external power source, deriving its energy from its suckers, which bathed in vats of balanced nutrients. The potato put on weight and tipped the scale at 60 kilos before Wilbur and Orville found the right balance in the nutrients to keep its weight stable. They named it P.O. Tate, which became the more informal Po as the twins got to know their creation better.

In order to give the world the benefit of their discoveries, the Spandau twins set up MonCul in the Arizona Desert. They chose the remote location because they wanted to keep their operation well away from population centres. They had no intention of spending time on the day-to-day operations of MonCul; they were far too busy. They appointed Heinrich Graft, a noted rocket scientist, as CEO and left it to him to set up the laboratories and hire the best staff that money could buy. The only contact Wilbur and Orville wanted was to be through a Chairman who would reside in Culver. The twins decided to be completely anonymous. Finding a Chairman was a problem. Wilbur wanted to hire a retired WO president, who was unfortunately arrested for corruption on the orders of his successor.

"We'll never find a suitable Chairman," said Orville. "We will have to reveal our identity and have Forbes and all the other magazines pestering us for copy. No way! I suggest Po for Chairman."

"Great idea," said Wilbur. "He is very discreet and has enough programming and computer power to knock the socks of any human manager. I'll inform Heinrich Graft that P.O.Tate is the Chairman."

The twins had grown so used to their garage, that they could not imagine doing their research anywhere else. They built an annex so that they could sleep there. They hired a famous architect to build a new home for their mother and developed computerised robotic systems to do the cleaning, cooking and ordering of groceries. This meant that they were not obliged to hire gossipy servants. When their mother complained that she was lonely, they arranged for her to have ballroom dancing lessons from D'Arcy Versey, a leading Hollywood dancer. The upshot was she ran off to Hollywood with the dancer and purchased a film company for him. It turned out to be a bad move.

For two years, Po ran MonCul impeccably, achieving an average bottom line of a billion. The company was floated on the NASDAQ, raising a record two trillion. Things started to go awry when Dr Richard Trickie was appointed CFO on a strong recommendation from Heinrich Graft. By this time, Wilbur and Orville had no reason to doubt Graft's honesty so they didn't ask Po to run a check. At the same time, Po was becoming erratic.

It all started when Wilbur was sitting at his desk, drinking a soda. He suddenly felt a sucker creep over his shoulder. He saw it dip into the soda, which it sucked up in seconds. Po became addicted to the caffeine in the drink. He sulked and stopped operating when the twins refused to fetch a fresh cold one. Soon Po was drinking a dozen bottles a day. When the twins were out, he ordered hamburgers and fries from McDuff's, putting them on Wilbur's credit card. When the Delivery Hound arrived, Po snatched the food and thrust it into the large mouth that he had secretly developed. When the Delivery Hound hung around for a tip, Po clipped it round the ear with his heaviest sucker and sent it away howling.

These circumstances combined to allow Trickie, Bach-Hand and Graft to milk the company and achieve a two billion loss. The trio spent their loot on enormous houses, yachts, and artwork. They held parties that cost millions. They became a byword for incredible extravagance. They knew that there would be a reckoning if they didn't take preventive action. They held a meeting in Sicily to lay their plans. They met at a remote villa on the south coast, which had once belonged to an Italian dictator. Its famous owner had fortified it like a medieval castle. Its current owners, the Parmesano family, used it for all their meetings and for receiving privileged guests.

One morning, the three conspirators were taking coffee on a terrace overlooking the ocean.

"Have you ever thought," said Bach-Hand, drinking in the blue Mediterranean panorama, "that none of us has ever seen the Chairman. You, Heinrich, were appointed by a head hunter, and you appointed us."

"We know he lives in Culver," said Trickie. "We could put a private eye on the case. We could consider bribing someone at Central Snoopers to make enquiries."

"And what would we do if we found him?" asked Bach-Hand, "Kill him?"

"Why not," said Trickie. "He's become very erratic lately. We would never have got away with what we are doing now, if he wasn't senile."

"Is it true," asked Bach-Hand, "that the company was started in a garage by two young Harvard graduates?"

"That's pure folklore," said Graft. "There's a

similar story about every big corporation in America. Let's leave the Chairman alone. If he's hit, someone competent might take over and we will be history. We need to sort out the smaller fry. The Stocking woman is a threat."

"What about Frank?" said Bach-Hand.

"I've seen the video of Frank lying shoeless in Acapulco," said Graft. "He didn't look much of a threat to me."

"What about the pigeon that Central Snoopers lost. It was watching Frank when it went out."

"Those birds are getting really unreliable," said Graft. "CS are losing three a week in San Francisco, more in New York. At a million a bird, they are probably due to be phased out."

"So," said Bach-Hand, "where is the Stocking woman?"

"She's living with her parents in San Francisco," said Graft. "She recently got a suspicious letter through ExpressCorp. When Central Snoopers checked her trashcan, they found a blank sheet of paper. The lab boys examined it and found traces of a chemical that caused ink to fade when exposed to the light."

"Highly suspicious," said Bach-Hand. "Shall we ask SolutionsCorp to eliminate her?"

"No," said Graft. "They have started to indulge in a little blackmail. You can only use them if you have nothing to hide."

"You can't trust anybody these days," said Trickie. "What is the world coming to?"

"One of us will have to do it," said Graft. "It shouldn't be too difficult. My father had a dozen ways of eliminating people."

"Was Stocking right, then?" said Bach-Hand. "Was he responsible for the Great African Massacre?"

"Of course not," said Graft. "You've heard of

Lemmings. It was a mass suicide."

At the Sicily meeting, it was decided that Graft would make the arrangements for eliminating Mary Stocking.

"I need to get back to MonCul," said Graft. "Why don't you two take a vacation? You've earned one."

"I'll stay here for another month," said Bach-Hand. "The food is wonderful."

"I'm thinking of popping over to Tangier before going home," said Trickie. "I hear the atmosphere there is very relaxing."

Heinrich and Trickie went to Tangier in the MonCul jet. The CFO dropped off and Heinrich returned to Arizona via various South American capitals where he contacted some of his father's friends.

Bach-Hand set about enjoying his vacation. He ate enormous meals, taking two dinners in the evening. He drank deeply of the delicious Sicilian wine. By the end of the month, his face was purple from overindulgence. The day before Bach-Hand was due to leave Sicily, Arthur Brown, the English doctor in Palermo, was summoned. When he arrived, he found the patient breathing his last.

"Heart attack, doctor?" asked old Luigi Parmesano, the feared head of the family.

"Suicide," said Dr Brown. "Suicide with a knife and fork."

Luigi's sense of humour was not well developed. "I said heart attack."

"Of course," said Brown. "I'll complete the death certificate and see to the formalities in Palermo. What shall I do about the funeral?"

Luigi smiled. Graft had left a large sum of

money for the expenses that Bach-Hand was about to incur. "I'll take care of the expenses," he said.

The very next day, there was a report that a man, carrying the passport of Dr Richard Trickie, had been stabbed to death, in unknown circumstances, in a back street of the Tangier Casbah. The police were not unduly surprised; people like Trickie often came to a sticky end. In the absence of anyone to claim the body, it was quickly buried in the European cemetery and the bill sent to MonCul. Luigi smiled.

Chapter Eight

Dr Frank greeted Mary Stocking with a hug. It was unusual for him to be demonstrative, but he was so relieved to see her.

"You look very well, Mary," he stammered.

"So do you, Frank," she said. "The last I heard of you, you were lying unconscious on a beach."

"A simple mugging, Mary," he said. "It can happen to anyone, and often does. Let me introduce my friends Coco and Radish.

"Hello, Mary," we said in chorus. "Are we glad to see you."

"And I'm glad to meet you two," said Mary. "Are you Dr Frank's creations?"

"Yes, we are. We're his children."

"Let me look at you, Radish," said Mary. "What are you made of and how are we going to recreate you?"

"I've no idea," I said. "Even Frank doesn't know how my Adamantine skin was produced."

"Is that true, Frank?" said Mary. 'You made Radish. You must remember what you did."

"I remember exactly," said Frank. "But I was never able to repeat the procedure successfully."

Mary looked at Frank. He was telling the truth.

"Was that you in the park?" she said.

"Yes," he said. "I make a pretty good woman, don't I?"

"You almost had me fooled," said Mary. "But, Frank, we are in a real fix. I think MonCul is going to silence me. And I'm pretty sure you are in danger, too."

Before Frank could speak, Coco said, "Hush. I'm intercepting a very interesting message coming from Luigi Parmesano in Palermo to Heinrich Graft in Arizona. Listen. I'll switch on my speaker."

We first heard a heavily accented Italian voice: "Yes, Heinrich, it's me, Luigi."

"Hello, Luigi. It's so nice to hear from you."

"I've got news for you: your friend Rupert died of a heart attack two days ago, and Dr Trickie was murdered in Tangier last night."

"Thanks, Luigi. I'm going out now. We shouldn't use the phone for these matters."

The transmission stopped.

"Well," said Mary. "What do you make of that?"

"It's incredible," said Frank. "It seems that Heinrich has eliminated the only two people who could incriminate him."

"Another message coming in!" said Coco. "Heinrich is on the phone to Central Snoopers. He wants a daily report on Mary Stocking's movements, where she goes, what she does, who she sees and who visits her. He's ordering a tap on all calls, e-mails and text messages. He wants Mary declared a terrorist."

"He's closing in," said Mary. "I'm afraid and I don't want to go back to my parents' house. But I must get a message to them. They'll be worried if they don't hear from me."

"I can do that," said Bernie, who had joined the

group after finishing his work.

"I can drop a maid off at your parents' house and she can pass a note. She'll explain that she got the wrong house."

"Thanks," said Mary. "But do you think I will be safe here?"

"No problem," said Bernie. "PoliceCorp never comes in here. We don't have TVs, radios or telephones. It's wonderful. We spend our leisure time on sports and the arts. We have a theatre, which doubles as an opera house and concert hall. We have a passable symphony orchestra, which gives twenty concerts a year. Tomorrow our opera company is putting on a performance of La Boheme. You must come along. We have libraries, an art gallery, a museum and a hundred clubs, which cater for every hobby you can imagine."

"That's incredible," said Mary. "I always

thought of the Under Class Areas as places of punishment."

"That was the intention," said Bernie, "but we run such a tight, self-sufficient, self- regulating community that outsiders are now trying to get in. The Federal Government won't allow it. That's why we have the fence. It's to keep people out."

The next day, Bernie set out to deliver workers to their employment. He dropped a maid at Mary's house. She rang the bell. Mary's father answered it.

"Did you order a maid?" she asked.

Mary's father looked at the minibus with the BUCA logo on the side. "No," he said. "You're mistaken."

"Sorry," said the maid. "I seem to have got the wrong house." She dropped a letter on the doorstep. "It's from Mary," she whispered and left.

If Bernie thought that his stratagem had succeeded, he was mistaken. The Stocking house was bugged in every room and Central Snoopers was able to relay Mr Stocking's conversation with his wife directly to Heinrich Graft. It was easy to guess from what was said that Mary was in the BUCA.

"Send a SWAT team in," Heinrich said. "Drag the woman out. She's a dangerous terrorist."

"Can't do that," said the duty officer, Miki Moto. "Our policy is to isolate the UCAs and leave them to their own devices. The last team we sent into BUCA, never came out. We heard they had been brainwashed and had joined the community."

Heinrich was depressed. Amazingly, he missed that weasel Trickie and the overbearing Rupert Bach-Hand. He informed the Chairman of the sad demise of his two colleagues and said that he would be acting CFO and Contracts Manager until new appointments were made. Graft had no intention of appointing anyone; he simply allocated their salaries to himself. There was a slightly crazed look in his eyes when he muttered, "It's me against the world. First, I'll deal with Contrary Mary."

A day later, Contrary Mary was drinking tea with Dr Frank. He was unabashedly smitten with the blond-haired, blue-eyed vision, who seemed to hang on his every word. Could it be that this Beauty was attracted to him, the Beast? He dismissed the thought and then wondered if a beautiful mind was sufficient to attract her, for

Frank had a really beautiful mind.

"Do you want to live in the BUCA permanently, Mary?" he asked. "You know that if you venture out you will be eliminated."

"Admirable though the BUCA experiment is," she said firmly, "I've no intention of spending the rest of my days here. I want to see my parents from time to time."

"Of course," said Frank soothingly. "Then we must plan our next move. Ladies first: what do you suggest?"

"I think we should find the Chairman and put our case to him. From what I gather, he is a brilliant manager. Surely he must see how unjustly we are being hounded."

"We can't assume," said Frank, "that he is an understanding, humane person. He may be like Graft, cold and heartless."

"It's a risk we have to take," said Mary. "Where

do we start?

"In Culver, Montana, the seat of the Chairman of MonCul," said Frank. "We'll have to go in disguise, of course."

"I expect he is surrounded by heavy security," said Mary. "We can't use electronic means to contact him, we'd be picked up at once. We'll have to try to get in to see him. Do you think it is possible, Frank?"

"There's only one way to find out," he said. "Let's get started."

The following week, before dawn on a cold Saturday morning, Bernie dropped a Mayan Indian and her daughter at the MagLev terminal. Frank, with his hooked nose, was a very convincing Indian.

"I think there must be some Native American blood in me," he said.

He looked at Mary who had been transformed by the BUCA Hospital specialists. She had been given a drug, which was normally used to treat albinism. It had changed her from blond bombshell to luscious Latin lady. His heart skipped a few beats every time he looked at her and she noticed.

"Frank," she said, "I know you are playing the part of my fond mother, but I know that you are not my mother, so cool it."

Frank blushed crimson under his wig and dark make-up. "Sorry, Mary," he said.

"And call me Maria," said Mary. "Our new ID cards say you are Teresa Martinez and I am your daughter, Maria."

Coco and I went along to handle electronic surveillance. Coco was in Frank's bag, but I was asked to grasp Mary's poncho and look pretty. I became a very attractive brooch.

The ExpressCorp MagLev train floated along the track at 200 kilometres an hour, arriving at Culver, just before lunch. Maria was hungry, so we went into McDuff's and ordered soyaburgers and fries. This suited vegetarian Frank, but Maria would have preferred a hamburger. She didn't dare to order one because they were horrendously expensive and no Indian could afford to eat beef. Frank, who had mastered Spanish while he was in Mexico, spoke to an old Hispanic woman sitting next to him in a high-pitched attempt at a woman's voice.

"Hello," said Frank, in Spanish. "Could you help me?"

The old woman looked at him suspiciously.

"You speak Portuguese?" she asked.

"Spanish," said Frank in heavily accented English. "I'm from Mexico. My daughter and I are maids. We heard we might get jobs with Mr Tate, the Chairman of MonCul."

"Don't know no Tate," she said. "I living here twenty year. Don't know no Mr Tate."

While Frank was wasting his time talking to the Brazilian woman, I was scanning the local telephone and radio traffic. Coco was probably doing the same, but he was shut away in Frank's bag. It didn't take me long to work out that Culver had seen better days. It appeared that it had been dependent on one textile factory, which supplied the state with working clothes. As fewer and fewer men and women did any physical work, demand fell away and the factory shut down. The town continued to provide the surrounding area with services, but the population had fallen drastically. I checked the computer files in the Municipal Offices and discovered that only one property paid big taxes. It was the Spandau estate in Wright Avenue. While I was waiting for Frank to finish his lunch I linked up with Coco and found that he had reached the same conclusion as me: a big boss like Tate would have a big house and the one at Wright Avenue was probably it. Frank was still talking to the old woman.

"You have very beautiful daughter," said the old woman.

"Yes," said Frank. "She's adorable."

I was almost dislodged from Mary's poncho when she elbowed Frank in the ribs.

We left McDuff's, and I told my friends what I had concluded. They walked to Wright Avenue, following my directions, and reached the Spandau house, which turned out to be all glass and aluminium and looked more like a huge space vehicle. Mary rang the bell at the gate and a janitor came out of his lodge."

"Hello," said Mary, making a better job of a Spanish accent then Frank. "Do you know if they want maids at the house? We are looking for work."

The janitor was very taken with Frank. What was it about him that attracted old men? He answered, ogling Frank.

"Nobody is living in the house," he said. "As it happens, I was looking for someone to do some dusting and general polishing and cleaning. I can't hire you, but I'll take you to see the owners. They're in their workshop."

The workshop was in the grounds of the big house. As we walked to it, the janitor chatted to Frank.

"My name's Juan," he grinned. "What's your name?"

Frank decided it was better if he didn't speak.

"You're the quiet one," Juan said. "You her sister?"

"Her Mama" croaked Frank.

"No!" he said. "You're too young to be her Mama."

We reached the workshop.

"This place used to be a garage," said the old man. "Wilbur and Orville made it into a workshop. They built an annex where they eat and sleep. They live here most of the time."

Frank could feel Coco jumping up and down in his bag. I tugged at Mary's poncho. I was picking up electronic traffic that was going mainly to the MonCul laboratories.

"Eureka," I whispered, just loud enough for Mary to hear.

Chapter Nine

D'Arcy Versey was busy on the sound stage at DV Pictures, filming a dance number for his first movie. He had already filmed the credits:

The D'Arcy Versey Story

Starring D'Arcy Versey

Directed by D'Arcy Versey

Musical arrangements by D'Arcy Versey

Choreography by D'Arcy Versey

Book and Lyrics by D'Arcy Versey

Produced by D'Arcy Versey and Gladys Spandau

The dancers, men in top hats and tails, were going through a routine, which involved a great deal of running up and down staircases, and D'Arcy was yelling abuse at them. He was in a bad temper because the last of his backers had pulled out.

"Cut!" he bellowed. "Get out of my sight the lot of you. I'm stopping the production. I'll let you know when I'm ready to start again."

He left the sound stage, and went into the office building, where his patron and erstwhile ballroom-dancing student was going through his books.

"Those are private, Gladys," he said.

"I'm entitled to know where my money has gone," said Gladys Spandau. "It looks as though you are going bust."

"It's all the fault of my backers," said D'Arcy. "The rats have run off."

"Left the sinking ship, you mean," said Gladys. "I must have been mad to buy this studio."

"Just give me another ten million and I'll complete the movie," pleaded D'Arcy.

"You're not getting another ten cents," said

Gladys. "If you think I am going to finance your new lifestyle, which includes intimate dinners with wannabe actresses, you are mistaken."

"What about your sons," he said in desperation. "They are rolling in it. I'll go and see them at the weekend."

"You can please yourself," said Gladys, "but I don't think Wilbur and Orville will be interested in investing in an old-fashioned musical about an egomaniac dancer. I'm going to find a buyer for this dump and move to Florida, well away from the movie business."

Gladys Spandau was as good as her word. She put the property in the hands of a real estate agent and left for Miami the next day. The agent did a deal with a property developer, who tore the studio down, built condominiums and shared the huge profits with the agent. If she had known about the scam, Gladys wouldn't have minded; she was pleased to see the back of the studio and the nauseating D'Arcy Versey. She couldn't imagine what she had seen in him. She had to admit, however, that he was a good ballroom dancer. She thought about visiting her sons in Culver, but decided against it. She sometimes thought they had forgotten who she was, they were so preoccupied with their research. She hated the house they had built for her; she hated the garage even more. She would start a new life in Florida with people of her own age, which was sixty-five.

D'Arcy Versey crept into his apartment via the fire escape to avoid the writ servers who were hanging round the entrance to his building. He selected the best of his clothes and packed a suitcase. He put some wads of hundred dollar notes in his briefcase, (his rainy day reserve), and left the building by the same route as he had entered it: it was one routine in which he never put a foot wrong. He hoped Gladys wouldn't get to her sons before he did. Knowing her, she would send them a message with two woollen scarves for Christmas, after she had knitted them. D'Arcy ignored his silver Mercedes, which wasn't paid for, and took a cab to the airport. He arrived in Culver as night was falling. He hated the dimly lit town, which was the antithesis of the bright lights of Hollywood. He decided on a quick con and a rapid exit.

D'Arcy took a cab to the Spandau estate, where Juan greeted him like an old friend. He shook Juan's hand warmly.

"I'm on my own," he said. "I came ahead to open up the house for Mrs Spandau. She'll follow in about a week."

"Good," said Juan. "I'll tell Orville and Wilbur that you are here."

"Don't bother them," said D'Arcy. "They will be busy in their workshop. I'll see them tomorrow. Which room shall I take?"

"You are in luck," said Juan. "We have two maids who are cleaning the house. All the rooms are ready. You can take the one next to theirs. I'll get Maria."

They went into the house, which was brightly lit. Juan found Maria and Teresa in the kitchen eating their supper.

"This is D'Arcy Versey, Mrs Spandau's friend," said Juan. "He has come to open up the house for her. He'll take the room next to yours."

Mary and Frank shook D'Arcy's hand and nodded a shy greeting.

"Would you like some supper?" Mary asked. "We have nut cutlets, spinach and some very nice salad. I'll show you to your room and then bring a tray up for you."

"Very kind," said D'Arcy. "And a bottle of the good red wine."

"Of course," said Mary, fluttering her eyelashes.

When Mary had returned from delivering D'Arcy's tray, Frank was in urgent conversation with Coco and me.

"What do you think of him?" said Frank.

"I was running a lie detector test on him as he was talking," I said. "Mrs Spandau is not coming."

"Then what's he up to?" asked Mary.

"I don't know yet," I said, "but I'll find out. I'll put a tap on his calls for a start."

The day we arrived at the Spandau estate, I

discovered that we were on the right track and that the garage contained the communications link between the Chairman and MonCul. I discreetly let Mary know and looked forward to seeing inside the building. This had to wait for another day, as a man of about forty, who turned out to be Wilbur Spandau, came out of the door carrying a crate of empty soda bottles. Juan told him why we were there, and he distractedly agreed to hire Maria and Teresa for a month to clean the house. He told Juan to give them a room. Juan led us to the house and showed us round. He explained that when the house was built, it was designed to be self-maintaining, but Mrs Spandau had left and the systems had been switched off. As a result the place had got very dusty. Mary was shocked when Juan said she and Teresa had to share a room. Frank was unconcerned. Mary thought of asking for two rooms, but decided it would look rather strange if she objected to sharing a room with her mother. Anyway, there were twin beds in the room and she knew Frank was a perfect gentleman.

We had been in the house for two days, when D'Arcy Versey arrived. Coco and I had been busy monitoring the communications with MonCul. It immediately became apparent that the Chairman was unwell. He frequently made mistakes in his transmissions; some of which were so garbled as to be incomprehensible. According to the traffic coming from MonCul, the laboratories were operating normally and operations had not been much affected by the sad demise of Trickie and Rupert Bach-Hand.

When D'Arcy moved in, the phone began to

hum. He bought an expensive diamond ring from a Hollywood jeweller using Mrs Spandau's credit card. Coco and I were sitting on the kitchen table while Maria and Teresa were having lunch when I intercepted an interesting call.

"Listen to this," I said, "D'Arcy is buying a one-way air ticket to Panama City, on Mrs Spandau's credit card."

"Why would he do that, when she is due to arrive here?" said Frank.

"I told you he was lying,"

"I wonder what he's up to," said Mary.

D'Arcy was up to no good. He invited Wilbur and Orville to lunch in the Culver Hilton. Over steaks, he was telling them about a great opportunity to make some money in the movie industry. He produced a sheaf of forged reports from non-existent previews, which lauded the D'Arcy Versey Story and said it was sure to get several Oscars.

"There's only one snag," said D'Arcy, "the studio needs a million and a half to make prints and organise distribution. If you invest, I can guarantee a twenty-five per cent return. I can't say fairer than that."

The brothers had only accepted D'Arcy's invitation to lunch because they felt that they had humour him to please their mother.

"I think we can do that," said Orville.

D'Arcy was overjoyed. He wished he had asked for five million, but it was too late.

"Could you make it a cashier's cheque," he said.

"I need to get moving quickly."

'I'll ring the bank this afternoon," said Orville.

Coco intercepted the call and told us about it.

"Now we know what D'Arcy is up to," said

Mary. "Should we do something about it?"

"Leave it to Coco and me," I said.

When Coco and I turned up at the garage that afternoon, he was carrying a stick over his shoulder on which hung a red kerchief in which I sat. He rapped on the door and Orville opened it.

"Hi!" Coco shouted.

Orville looked down and, strange to say, he wasn't a bit surprised to see a coconut with limbs, one Cyclops eye and tombstone teeth standing on his doorstep. It showed how far science had progressed in the direction of utter insanity.

"Hi," he said. "What can I do for you?"

"Can we come in?" said Coco.

"We?" he said.

I piped up. "Hello, I'm Radish and his name is

Coco. We are MonCul products."

"I'm sorry," said Orville, "You can't come into the lab. I'll get Wilbur. We'll go over to the house."

We went into the library.

"Do you two partake of refreshment or are you completely autonomous and self-sufficient? asked Wilbur.

"We're fine," I said.

At that point, D'Arcy walked into the room. Coco and I closed down.

"Do you have my cheque?" he asked.

D'Arcy's fingers were about to close on the cheque when Coco's eye and mouth flipped open.

"Hold it there!" he said. "You, D'Arcy Versey, are a con-man. Mrs Spandau is not coming here; she's in Miami and intends to stay there. You have obtained a vastly expensive diamond ring on her credit card. No doubt the ring is intended for the Panamanian actress you are going to join using the one-way air ticket you have just purchased, also on Mrs Spandau's credit card."

"Is all that true?" asked Orville.

"How can you believe the word of a coconut?" said D'Arcy. "It's all lies."

"I can vouch for the truth of it," I said. "What is more, production of the D'Arcy Versey Story has been abandoned, only the credits are in the can, and the studio is up for sale."

D'Arcy picked up a heavy bookend and brought it down on me. "You lying radish," he screamed. "Take that!"

I flipped the bookend off me and raised myself to my full height.

"I have Mrs Spandau on the line," I said. "She will back me up."

I switched on my speaker and Gladys Spandau

spake.

"Hello, boys. Mr Radish tells me that D'Arcy, the crook, has been using my credit card and is in the process of conning you out of a million and a half. Get the ring. I'm not going to let some Panamanian floozy have it. Don't give him the cheque, but let him keep the air ticket. Good riddance to bad rubbish. Do give him the boot. Bye. I'm knitting you scarves for Christmas."

"Well!" said Orville. "That was just like Mummy, not letting us get a word in. As for you, Mr Versey, you can leave immediately."

He picked up the phone.

"Juan? Come over to the library. Mr Versey is leaving."

Juan arrived. "Yes, Mr Orville?"

"Conduct Mr Versey to his room. Give him five minutes to pack. A taxi will be here in ten."

Juan watched D'Arcy pack. "I knew you was a crook," he said. "I saw you looking at the hallmarks on the silver. What are those?"

"My mother's candelabras," said D'Arcy. "They're family heirlooms."

"Take them out," said Juan. "They came from our dining room. You took anything else?"

"Take the blasted candelabras," said D'Arcy petulantly, hoping that Juan had not noticed the disappearance of a silver cigarette box from the living room.

"And you can leave the silver box," said Juan.

D'Arcy flung the box on the bed.

"And the silver cutlery," said Juan.

D'Arcy Versey was about to board the plane for Panama when two PoliceCorp detectives closed in and snapped handcuffs on him. From the information supplied by Coco, they had enough on D'Arcy to put him behind bars. In jail, he put on some stupendous musical productions from the early 20th century: he was in his element.

Chapter Ten

Heinrich Graft stalked the MonCul labs with an insane look in his eye. The staff kept their eyes down. If an unfortunate employee glanced at the mad CEO, he or she was terminated on the spot.

"I want discipline!" screamed Heinrich. "How can I create the New World Order if you don't respect your leader?"

He went back to his office where a small, swarthy man, wearing reflecting shades, was waiting for him.

"Hello, Guido," he said. "How is Luigi?"

"My father is well and sends his respects," said Guido. "Who's the hit?"

Graft threw a photo on his desk.

"Name of Mary Stocking," he said. "Left-wing

agitator and terrorist."

"She's history. The money?"

"In Zurich."

"Ciao."

"Ciao."

Graft was depressed. He pressed a button and the daily reports came up on a screen. What he read depressed him more. The corn crop had failed in Chiapas. At first, he thought the farmers must have planted illegal seed, but he was wrong. They had bought MonCul seed and the Doomsday gene had kicked in before the corn matured. He sent for his chief scientist, Krull.

"What's this about Chiapas?" he said.

"It's not just Mexico," said Krull. "I'm getting reports of crop failures from Central and South America. If this continues, the North American farmers will make a fortune when the price goes up."

"So it was deliberate," said Graft.

"No," said Krull. "Just fortunate for our farmers. I've suspected for a long time that the doomsday gene was unreliable and now I have been proved right."

"Are you being deliberately stupid or does it come naturally?" said Graft. "Millions of Hispanics are going to starve and you are gloating about the rise in the price of corn. We can't get away with this. Get the Contracts Manager to buy up as many corn futures as he can lay his hands on. I want us to be able to supply the farmers who are in trouble and make large profits at the same time."

"You're the Contracts Manager," said Krull.

"So I am," said Graft. "I'll close the Corporate Governance and Ethics Department and get Angela to do it. You can go. And, by the way, you are terminated."

The news of the impending disaster was not sent to Culver, but Coco and I were completely *au fait* with the facts. He and I were staying openly in the Spandau house and were able to converse with Maria and Teresa, even when Juan was around, and Juan was around a lot. Every day he brought flowers from the garden for Teresa, and frequently proposed marriage. Teresa protested vehemently that she was a respectable widow who had promised her husband on his deathbed that she would never remarry.

Orville and Wilbur often ate in our kitchen now because Maria was such a good cook and Wilbur had taken a fancy to her. Coco and I had not yet been inside the garage, but we reported the latest events to the twins when they were eating lunch or dinner.

"Boys," said Coco, "you're being kept out of the loop. Graft is buying up corn futures and the crop is failing in the south. The authorities are gearing up for tortilla riots."

"We have a problem," said Orville. "The Chairman is not well and he handles all the MonCul business. Wilbur and I are doing our best to sort things out, but we're not making much progress."

"If there is anything we can do," I said, "just ask."

"Thanks, Radish," said Orville. "You can help us by monitoring everything that comes out of MonCul."

"Here's one for you," said Coco suddenly. "A Guido Parmesano is talking to his father, Luigi, the big cheese of the clan. That's the family Rupert Bach-Hand was staying with when he died. Apparently Guido has taken a contract and is asking his father's permission to contact the New York branch to recruit two soldiers to help with tracing the target."

"What has that got to do with us? asked Orville.

"It's got everything to do with you," said Coco, "Guido Parmesano was in Heinrich Graft's appointments diary. They had a meeting recently."

"You mean to say," said Wilbur, "that someone connected with MonCul is the target?"

"Obviously," said Coco.

Wilbur looked at Maria who had uttered a sigh. He often looked at her.

"Maria," he said solicitously, "are you all right? You are looking very pale recently."

"I'm looking pale," said Maria, "because the drug is wearing off."

Wilbur was alarmed. "Drug?" he said. "What

drug?"

"The one that changed me from a blonde bombshell into a luscious Latin lady," she said. "I'd better come clean. I came here to appeal to the Chairman for protection against that madman, Graft."

Wilbur and Orville listened with open mouths as she made a full confession. When she had finished, there was a long silence. Wilbur spoke first.

"It's a lot to take in at once," he said. "It seems we are in deep trouble. Not only is the Chairman gaga, but Graft is an assassin and Moncul is rapidly going down the tubes. I can assure you that the Chairman did not know anything about your problems, Maria, or should I call you Mary? He was told that you had resigned to spend more time with your family. And Teresa, or should I call you Dr Frank? I have been aware for some time that you didn't behave like a maid. I saw you doing the Times crossword on the kitchen table. You completed it in five minutes. You can now drop your disguise."

"I'd better talk to Juan," said Mary. "He'll be upset."

"Be gentle with him," said Frank. "I'm really very fond of him."

"Since it is confession time," said Wilbur, "you'd better all come and meet the Chairman."

We all trooped across to the garage. Wilbur unlocked the door and ushered us in. The horrible sight that met our gaze was fascinating in its repugnance. The Chairman was a huge potato. One sucker was dipping into a soda bottle and two were thrusting beef-burgers into a voracious mouth. Frank was fascinated.

"Wonderful!" he exclaimed. "You have

developed a potato to run a multi-billion dollar industry.

"It's not all that revolutionary," said Orville. "A vegetable marrow was the head of the WO for years before he became the main course at a vegetarian dinner. Every one of our presidents since, and including, Dubya, has been a vegetable. Heads of state and, even the heads of the World Organisation, don't make decisions. The Oligarchs behind them have the real power. Po, our chairman, was different. He made MonCul run like clockwork. But he went the way of all flesh and developed bad habits. He is addicted to junk food. He gets fatter every day and is losing all sense of proportion in his addiction. I tried to persuade him to cut his intake to three hamburgers and six sodas a day. That would have stabilised his weight. He wouldn't have any of it. He shut down if we didn't keep a constant supply of burgers, soda and the occasional apple pie coming. He now weighs two hundred and fifty pounds, whereas his optimum weight is a quarter of that."

"Why don't you replace him?" said Frank.

Po swung his heaviest sucker and just missed taking Frank's head off.

"See that? said Orville. "He used to be a sweet potato. Now he's an aggressive spud."

Po finished his soda and threw the bottle at Orville. In his obese clumsiness, he was unable to control his aim. His suckers flopped down and he visibly sagged into a mass of carbohydrate.

"Now he's shut down," said Wilbur. "He'll be out until the withdrawal symptoms kick in. Then he'll get on the phone to McDuff's for a take-out."

Frank repeated his question.

"We are trying to replace him," said Wilbur,

"but we have been adding nano-computers and programming to him for years and not keeping proper notes. We have started on creating a replacement, using a turnip, but it will take years to finish."

"Well, look no further," said Coco. "Radish and I have all the nano-circuits and programming you need to run a company. We don't ingest organic products so we are not likely to get addicted to caffeine or any other stimulant."

"Coco is right," said Frank. "He and Radish would make a perfect team."

"Wait a moment," I said. "Coco volunteered to be Chairman, not me. I have no desire to be a slave to a capitalist venture, which makes billions out of the misery of poor farmers who can no longer plant their own seed. I find the MonCul operation quite immoral. I intend to start a crusade to end the abuse of flora and fauna in the mad race to genetically modify every living species."

Mary, Frank, Wilbur and Orville stared at me.

"I understand, Radish," said Coco. "You have ethical programming, which I lack. I'll enjoy being a big boss. It's not every day a nut gets to control the fortunes of a multinational company."

"Look!" said Mary. "Something is happening to Po!"

We all stared in horror as Po began to disintegrate and liquefy into rivulets of corruption. The smell of a rotten potato is well known to you, Dear Reader. Please don't let your attention wander. There's nothing on the box and I advise against getting a snack because the description about to follow will put you off your food for at least a few hours. Mary screamed and ran out of the garage. As scientists, Frank and the twins had to stay to observe Po's end. We watched with disgust as a quarter tonne of putridity oozed and bubbled, releasing noxious gases that defied description in their offence to our olfactory organs. Yellow waves carried larger bits of circuitry and motors out of the garage. Po's emptying skin sagged and his many eyes blinked spontaneously; his suckers twitched, twisted and writhed. I could not resist quoting a French Romantic dramatist. I declaimed in translation because my companions, except for Coco, knew no French. If they had, they would never have been able to say MonCul without a titter.

'The world is nothing but a great sewer in which seals scramble and twist on mountains of slime.'

So wrote Alfred de Musset. But I digress. The defunct potato heaved and ventilated. Wilbur, the more sensitive of the twins, retched and ran out of the garage. Orville and Frank stayed to the bitter end. Po's skin, voided of its putridity, filled with gases and then exploded in a shower of disgusting fragments. We staggered out of the garage.

"I suppose Po's death was for the best," said Orville philosophically. "I was thinking of poisoning him anyway."

"We can't use the garage anymore," said Wilbur. "What are we going to do with it?"

"I'll get Juan to salvage what he can from the annex," said Orville. "Then I'll torch the whole place. To think that in future years it might have become a place of pilgrimage for MBA graduates about to start their own business."

"We can erect a plaque," said Wilbur. "You know the sort of thing, "In this garage...."

"I filmed the whole event," I said quietly. "I intend to beam it to all the schools in the world to demonstrate the dangers of junk food. If you have any shares in McDuff's, Wilbur and Orville, get rid of them; they will be junk too."

I really was becoming a self-righteous prig.

Chapter Eleven

Guido Parmesano was in conference in San Francisco with two soldiers from the New York chapter of the family.

"You guys picked up the trail yet?" said Guido.

"Sure, Guido," said Carlo, the tall one of the pair. "My friend, Anthony here, has been into Central Snoopers and talked to his brother."

Anthony, the short one of the pair, made his report. He didn't use notes because he was illiterate.

"The Stocking woman is with mother. She staying at a house in Culver, belong twin nerds, Wilbur and Orville Spandau. A Delivery Hound what works for McDuff's, say like brothers are addicts to beef-burgers and soda, also addict is a potato who live with them."

"Wait a minute," said Guido. "You telling me the Spandaus are cohabiting with a potato? That's disgusting. It's against nature."

"They just good frens," said Anthony.

"Anthony, how long have you lived here?" asked Guido.

"All my life," replied Anthony. "I born here."

"Then why is your English so bad?" said Guido.

"I speak Englis good," said Anthony. "What for you critic me?"

"Never mind," said Guido. Here's the plan. We drive to Culver and stake out the house. I have a sniper rifle with a night sight. When the Stocking woman shows, she's history."

Carlo and Anthony took turns to drive to Culver,

where the trio settled down on a grassy knoll, overlooking the Spandau house. They could see the pile of blackened ashes that used to be the garage.

"Look like they torch that," said Anthony. "Maybe insurance job."

They waited in vain for two nights without spotting Mary Stocking. They frequently saw the two maids who worked at the house, but never the blond target. On the third night, Guido said, "I can't wait about any longer, I'm going in." He pulled a ski mask over his face. "Come with me Anthony," he said. "You stay with the car, Carlo. I'll call you on the mobile when the job is done. Drive up to the front door and pick us up. If you see anything we should know about, use your mobile."

In the house, Mary and Frank were in the kitchen with Coco and me. Wilbur and Orville were in San Francisco, buying new lab equipment. The kitchen door, which wasn't locked, opened and two masked men came in.

"Hands on the table," said Guido, pointing his rifle at Frank's head.

Frank and Mary did as they were told.

"Where's Mary Stocking?" said Guido, throwing her photo on the table.

Mary responded without too much thought, "She dead. She burn in fire."

"Which fire?" said Guido.

"No speak English good," said Mary, laying on a Spanish accent with a trowel. "Fire outside."

Anthony was furious.

"You people," he said, "you comes and takes our job and no spik Englis. Where Stocking?"

Mary and Frank shook their heads.

"Go through the house, Anthony," said Guido. "If you find Stocking, shoot her."

Anthony waved his pistol. "Right, Guido," he said. "She history." He returned five minutes later. "Nothing, zilch, nada, sweet Fanny Adams" he said, demonstrating his command of synonyms.

In the meantime, both Coco and I were transmitting to the Culver PoliceCorp office. We gave detailed descriptions of the two men, (except for their faces), and when Guido called a third member of the gang, telling him to drive the Mercedes to the front door, we added that information. I monitored the radio traffic between the Black Death helicopter and the ground force. Then I heard the dull sound of the silent-running helicopter as it hovered overhead. The next message said that the ground force had intercepted a man driving a Mercedes and arrested him. A searchlight flooded the kitchen with blinding light. Coco tweaked up his speaker to maximum and bellowed, "Drop your guns! Don't move or I fire. Pick up their guns, ladies."

Mary and Frank picked up the guns and trained them on the gangsters. Guido turned his head and looked bemused when he saw no-one.

Move out!" said Coco, pointing to the door and fluttering his eye and mouth flaps. Guido's eyes popped, but he went out like a lamb into the hands of the waiting PoliceCorp troops.

We congratulated Coco on his quick thinking, which had ended a very dangerous situation.

"It was nothing," said Coco. "We captains of industry are used to thinking on our feet."

The outcome of the arrests was most unsatisfactory. Luigi Parmesano had some influential friends at Central Snoopers. They persuaded PoliceCorp that the three Mafia men were their employees who had been on a mission to arrest Mary Stocking, a dangerous terrorist. The three were released. Guido went home, and Carlo and Anthony were given jobs at Central Snoopers.

When the twins returned, they were appalled by the news that Mary had been the target of assassins.

"It looks as though you will have to keep wearing your wig, Frank," said Wilbur. "And you'll have to keep taking the pills, Mary. Just until we sort out that madman, Graft. I'll get in touch with PoliceCorp. We have enough evidence of corruption to have him arrested."

When a PolicCorp squad went to MonCul, Heinrich Graft had disappeared. He hadn't fled to Vanuatu. He was walking through the desert, heading for the Sierra Madre. Dressed in rough working clothes, his hair and beard unkempt under a Mexican sombrero, he was leading a pair of donkeys laden with gold prospecting gear.

"I know there is gold in the Sierra Madre," he muttered to himself. "I saw an old map, or maybe an old movie."

When night fell, he made camp, cooked bacon and beans and brewed coffee. He realised he was happy. He smiled at his donkeys, which were munching hay.

"This is the life," said Graft. "Free of all the nonsense and trappings of big business."

He unrolled his sleeping bag, crept in and dropped immediately into a deep sleep.

The next morning, he woke early and scanned the horizon. When he saw a cloud of dust approaching, he sighed with relief. The SUV stopped and the driver got out.

"Hi," he said. "I'm Mike from Supplies-R-Us. Are your Gabby Bogart?"

"Sure, I am," said Graft. "You got my supplies?"

"I have," said Mike. "Here's your box."

He took a box containing cans of beans, bacon, coffee and sourdough from the SUV. There was also a bag of feed for the donkeys.

"There's a fresh battery for your satellite phone in the box. I'll take yours for recharging. Is your GPS working?" said Mike.

"You found me, didn't you?" said Graft.

"My orders say you are due for a weekly delivery," said Mike. "Are you sure bacon and beans is a suitable diet for a prospector?"

"You bet," said Graft.

"I'll refill your water bottles," said Mike. "You want distilled or mineral water?"

"Whatever," said Graft, getting into the part. "I ain't fussed. I ain't one of your namby-pamby city folk."

"I have to tell you," said Mike, "that if you go into the hills, we'll have to use a helicopter drop."

"There's gold in them thar hills," said Graft.

"Can you afford a helicopter drop?" said Mike. "It will cost major bucks."

"Lookee here, pardner," said Graft. "I got one hundred and eighty billion bucks in my account, and you ask me if I can afford a helicopter drop."

"Sorry," said Mike, who was unfazed by Graft. The latest craze among the super-rich was gold prospecting the hard way, with shovel and pick and a donkey or two. Supplies-R-Us was specialising in deliveries to them and making a nice profit. Unlike Gabby Bogart, most of Mike's customers ordered the finest foods to compensate for their hardships.

"Let me have your co-ordinates when you want your next delivery," said Mike, climbing aboard his SUV. "Bye, Gabby."

"Adios, amigo," said Graft. "See ya next week."

Back in Culver, Coco was working flat out on the affairs of MonCul. He soon discovered that the WO advance for R and D on Adamantine had been misappropriated and was sitting in a bank on a ship cruising in international waters.

"It's the latest scam," Coco told the twins. "These floating banks are hard to deal with. I'll track this one, and when it puts into port to re-supply, I'll get InterPolCorp to seize it."

In the meantime, Coco informed the WO of the theft. He also sent a report, stating that the claims made for Adamantine were false and that MonCul was withdrawing from the contract. Development of the invisible bomber went ahead on the grounds that, if it was invisible, it would not need an Adamantine skin. The failure of the corn harvest in the south was rapidly dealt with. Coco ordered free deliveries to the affected areas to start as soon as the northern harvest was in. Eventually, there was a surplus and, because Angela had bought up most of the futures, it was sold off at a huge profit to countries that could afford it.

I was a little piqued because Coco had very little time for me. He had set up a state-of-the-art office and conference room and had frequent meetings with the twins, which I sat in on, not saying much, because I had nothing to say.

"I have been working on a strategy for MonCul's future," said Coco at one meeting. "I propose privatising the shares, which have sunk to practically nothing, and turning the company into a charitable foundation."

"That sounds good to me," said Wilbur, whose interest in MonCul was evaporating. "What then?"

"We get out of GM foods and into pharmaceuticals. The world is crying out for cures for malaria and other diseases, which kill millions of people every year. We develop medicines and supply them at cost to poor countries."

"Sounds good," said Orville.

I drifted off into the ether and tuned into a satellite, which was surveying the southern deserts. Suddenly, I caught sight of a familiar figure. In spite of his beard and long hair, I recognised him by his voice.

"I know there is gold here," he was saying. "And I'm keeping it. You two had better watch out. I know you. You are only waiting for me to dig it up and then you'll kill me."

Coco zoomed out to see who Graft was talking to. There was only a pair of donkeys in view. Graft was talking to them, while he dug furiously into the sand. I put the pictures up on the conference room screen.

"Take a look at this," I said. "These pictures are coming from a MineralCorp satellite. It keeps track of prospecting activities. It has spotted an unlicensed digger, by the name of Gabby Bogart. Do you recognise him?

"It's Graft!" said Frank. "But look at the readings on the minerals scale. There really is gold in that location."

"And MineralCorp know about it," I said. "That's why the satellite is staying on him."

Suddenly Gabby lifted a heavy chunk of ore. "What did I tell you," he said to the donkeys. "Gold!" He fell to his knees hugging the nugget, which looked as though it was worth a fortune. He staggered to his feet and loaded it into a basket on one of the donkeys, and moved off into a cave where the satellite lost him.

"That's all very amusing," said Coco, "but we have important matters to discuss. We have agreed to the change in our commercial activities, now we need a change of name."

"Why? said Wilbur. "It's a perfectly good name, which Orville and I chose."

"In the first place," said Coco patiently, "after the excesses of the previous management, the name of the company is mud. In the second place, in French, it means 'MyAss' and that doesn't refer to the kind of animal Gabby Bogart is using."

"In that case," said Orville, "what do you suggest?"

"CoCoCo," said Coco. "It has a nice ring to it."

I stopped paying attention to the meeting and intercepted signals coming from Gabby Bogart. He was on his satellite phone, asking Supplies-R-Us for an immediate helicopter rescue. He stressed that his life was being threatened by two desperados, who were trying to steal his gold. Within half an hour a Funky Chicken helicopter was on its way. When it landed, Gabby loaded a heavy bag into the helicopter and jumped aboard.

"That will teach you, you varmints," he yelled at the donkeys. "You thought you could get the better of Gabby Bogart. Well, you can't, so there!"

Six months later, he was guiding a two-hundred-year-old steamer up the Amazon. His only passenger, a nun from the Amazonas convent, was not happy.

"Mr Bogart," she said, "if you persist in drinking the river water, you are likely to come down with a fatal ailment."

"Don't worry, Sister," he said. "It takes more than amoeba to get the better of Gabby Bogart. Anyway, I drink it half and half with my water purifier, manufactured according to an ancient Scotch remedy."

The romantic river trip ended with the nun asking Gabby to join her at her jungle mission. He rapidly became a living legend in Amazonas. He had taken to playing the cello, which he had studied as a child. He and the nun ran a clinic in which they treated the Indians for a variety of minor ailments. He was being tipped for a Nobel Prize when things turned sour. A visiting missionary discovered that Gabby was trying out experiments, which terminated in the termination of the patient. Gabby disappeared before BrazPolCorp could arrest him. The nun was inconsolable. But I leap ahead of myself, which, you must admit, Dear Reader, is a difficult thing to do.

Mary Stocking was relieved when Gabby left for Brazil. She pondered on the stupidity of a man who was about to have her killed just because he thought she could reveal the scam connected with Adamantine. It struck her that most corrupt men, she couldn't think of a female example, were far too ready to break the law to protect themselves, even when it was unnecessary. She thought about the American president who had organised a burglary to increase his chances of re-election, resulting in his ignominious resignation.

Mary was happy that she had Wilbur as a friend. He comforted her when she was depressed, brought her flowers and took her out to dine expensively, but never once suggested that they might become a couple: the useless idiot. One evening, she and Wilbur were drinking coffee in front of a blazing fire in the Spandau living room. She put her cup down, took his cup from his trembling fingers, put him in a half-Nelson and said, "What are you, Wilbur Spandau, man or mouse? When are you going to take me in your arms and say you can't live without me?"

She released Wilbur, who said, "But I can live without you."

"Wrong answer," said Mary, putting him in a headlock. "We are getting married next week and we are going to produce heirs to benefit from the billions that you and Orville have in your accounts. How can you contemplate the prospect of all that money going to a distant cousin in Milwaukee?"

"You have a point," said Wilbur. "Will you be my wife?"

Mary's parents and Gladys Spandau came for the wedding, of course, and Frank was the best man. Mary insisted on a long train, which Coco and I had to carry. Since our legs were rather short, we had difficulty in bearing the train at a suitable speed. Consequently we were dragged to the front of the Universal Vegetarian Church and Restaurant where Wilbur and Mary exchanged vows.

"Do you promise to give this woman all your money?

"I do."

"Do you promise to abstain from meat till death do you part?"

"I do."

"I now pronounce you man and wife."

That was it. Many in the Church had tears in their eyes, but dried them when the chef, who had conducted the ceremony, said he was providing meat dishes for non-conformists at the wedding feast. He was proud to announce that the WO had donated a whole ox to be roasted on a spit, an event, which had not happened for a decade. It was quite a wedding. Frank was very brave in the circumstances. He had long had an unrequited passion for Mary. His luck was in at the wedding, however, because one of the bridesmaids, Angela from CoCoCo, got intoxicated and decided that Frank resembled a long-dead body-builder governor of California who also had tombstone teeth. She married Frank two months later.

Chapter Twelve

Crab, you remember Crab, Dear Reader, the GM Hermit who had been transmogrified into a Relay Station for WorldCom Galactic Communications by MonCul, now CoCoCo, was sifting sand, when a submarine surfaced off her Panamanian beach. Crab took notice, and whistled to her adopted son and daughter to come into her shell. Crab was very happy with her children. She had been given the Doomsday gene when she was modified which meant that she could never have eggs. She had become so desperate that she stole two from a neighbour and nurtured them into her beautiful babies. The neighbour didn't notice because she had about a million eggs.

Crab watched a truck roll onto the beach and unload a large pile of cartons. This was not the first time a truck had made a delivery. Crab was aware that her children were in danger of getting squashed so she made sure they were always safely inside her two-roomed shell. A small boat ferried the cartons to the submarine. During the loading process, a carton was dropped on the rocks and burst open. A gross of lollipops spilled into the sea and bobbed in the surf. Once the cargo was on board, the submarine made for the open ocean and submerged.

The next day, there was a commotion on the beach. Crab gave her babies strict instructions to stay in their shell and went to investigate. A crowd of Fiddler Crabs was attacking a lollipop, which had been washed up minus its wrapper. The Fiddlers, who were a very nice family, had gone crazy. They were trampling on each other and swinging their claw to get to the lollipop. Normally, the Fiddlers never touched anything other than what the sea provided by way of small fishy morsels. Now they were eating candy and fighting each other to get at it. Something was wrong and Crab was worried. This was no environment in which to bring up children.

Since her defection, Crab had continued to listen to communications on the WorldCom Galactic network, but had never transmitted anything because she was afraid of being tracked down and eliminated as a deserter. Day after day, the lollipops washed up on the beach and the Fiddler riots continued. Crab took a risk and contacted me in Culver. The message was short: 'Call the Panamanian Crab'. I knew at once what was required and got in touch with Crab with: 'Sit tight, Radish on way. I repeat, Radish on way.'

Frank and Angela had moved to Los Angeles after their wedding. Frank was teaching at UCLA and had taken up sailing as a hobby. Angela was also a keen sailor, and, every weekend, the happy couple was out on the ocean breathing fresh air, rather than the LA smog. It was on one of their nautical excursions, that they got my call.

"Frank," I said, "Radish here. I must get to Panama a.s.a.p. Crab is in trouble. Can you help?"

"Well," said Frank, "as it happens, we start our long vacation next Monday. We can provision up and get you down to Panama. Can you be here for then?"

"No problem, amigo," I said. "Wilbur and Mary will bring me down and see us off."

Our group, when it arrived at Frank's bungalow on the UCLA campus, numbered five: Wilbur,

Mary, Orville, Coco and me. We were all intent on taking a cruise to Panama. Coco left the running of CoCoCo in the tendrils of a cucumber, named Q, which the twins had developed into a competent assistant for him.

"The sign of a good manager is his willingness to delegate," said Coco. "A good executive should also have lots of time to spare if he is functioning well. I have tried golf, but it's not my bag. Sailing, on the other hand, is right up my street."

The evening before we sailed, we had a brainstorming session. What were we to make of Crab's appeal for help? Coco and I had been concentrating on monitoring the electronic traffic coming out of Panama. The latest President had taken power in a coup engineered by Central Snoopers. He called himself General von Rouse, and only ever appeared in public driving an ancient Tiger Tank. I had decoded some messages, which turned out to be large orders for lollipops to be smuggled into LA and San Francisco by submarine.

"There's something wrong with your decoder," said Wilbur. "Whoever heard of a clandestine trade in lollipops? It must be code for drugs."

"Wait a moment," said Angela, Frank's young, beautiful and talented wife. "Don't you remember the burger riots? They were caused by a taste enhancer in Superburgers. It was even tried in soda and ice cream. It was addictive."

"That was a McDuff experiment that went badly wrong," said Orville. "Their labs developed an ingredient derived from monosodium glutamate, or MSG. Strictly speaking it wasn't a narcotic but it had similarly addictive properties."

"McDuff's turnover went through the roof," said Angela. "People couldn't get enough of the

Superburgers. Muggers were attacking customers leaving outlets, snatching their burgers and ignoring Rolexes and fat wallets. The normally passive Bureau of Firearms and Food had to act when pet dogs started to raid McDuff's."

"I remember that," said Mary. "Some owners tossed bits of burger to their dogs and they became addicted. Droves of dachshunds, platoons of poodles, battalions of beagles and worst of all, serried ranks of rottweilers besieged the stores. Tri-monosodium glutamate, or TMSG, was classified a dangerous drug and production was stopped."

We anchored in the bay off the beach where Crab had her family home. My thoughts went back to the day when Coco, Crab and I had landed in out reed boat. I thought about the progress Coco and I had made, Coco to the first rank of the business world, and myself to the status of a learned radish with the knowledge of several millennia stored in my nano-memory. I had a word with my friends.

"I believe," I said, "that it would be better if Coco and I went ashore to interview our friend, Crab. The situation in Panama is, as usual, insecure. If there are any badly intentioned individuals observing us, they will not be aware of a coconut and a radish, washing up on the beach."

Our friends were in agreement, so I clung to Coco's antenna when he jumped into the sea and paddled away with his feet and arms to the beach. We were greeted by Crab's whistle. She rushed to us with her two children and let Coco hug her.

"I'm not toxic anymore," I said, and got a fishy

hug.

We retreated rapidly to the lee of a large rock.

"So, Crab," I said, "what's the problem?"

"The problem," said Crab, "is that the Fiddlers, formerly such a nice family, who always waved their claw when I met them, have now become a mob of hooligans, with no time for social niceties. The only thing they can think about is getting their claw into the next lollipop."

Coco and I must have looked puzzled, because Crab continued.

"Somebody is shipping large consignments of lollipops by submarine to LA and San Francisco and some fell in the ocean. That much I know from my listening activities, which are much reduced, owing to the necessity of taking care of two bouncing babies. But I have been in a position to see what happens to Fiddler Crabs when they get a taste of one of the lollipops.

They go insane! They become crazed addicts."

"Can we get hold of an example for testing?" said Coco.

"Fat chance," said Crab. "They're devoured as fast as they wash up on the beach."

As she spoke, a red, plastic-wrapped candy on a stick was washed to our feet.

"Grab it and run!" said Crab. "The Fiddlers are coming!"

Coco grabbed the lollipop, I grabbed hold of his antenna, and he dashed into the surf just as a band of crazed Fiddlers came into view."

We struggled back to the boat, or Coco did, and we were hauled back on board.

"It's horrible," said Coco, brandishing the lollipop. "These things are destroying the Fiddlers. Once they were happy, harvesting the natural produce of the ocean. Now they can only think of when the next fix is coming in on the tide."

"Let me have that," said Frank. "I have brought my ChemScan set along. I can test for MSG and the derivative TMSG."

Before he could test anything for anything, a gunboat leapt into view and came broadside alongside our boat showering us with a great wave of spray.

"Show-offs!" said Wilbur. "Why can't they do things with a touch of decorum?"

Good manners and our visitors were total strangers. Three sailors boarded us in a flurry of waved pistols, and slammed handcuffs on Orville, Wilbur, Mary, Frank and Angela. They didn't see any threat in a coconut and a radish. Big mistake!

Our vessel was attached to the gunboat and towed to a jetty in Panama City. I guessed that my friends, except for Coco, would be taken ashore, so I discreetly clambered up Mary's sleeve and attached myself to her blouse, once again assuming the appearance of a pretty, if somewhat unusual, brooch. We were hustled ashore and loaded onto two jeeps, which sped through the city to a factory complex. Over the gate, was a slogan, 'TMSG MAKES YOU FREE'.

"What do you think of that?" said Frank.

"It explains what is in the lollipops," said Mary. "But I can't see any of the local people with sticks projecting from their mouth."

"Perhaps they are all reserved for export," said Orville.

We were soon enlightened. The brass plate on the door of the building we were propelled into bore the name Dr Gerbil Sneed, Chief Executive Officer.

"There can't be anyone else with a name like

that," said Frank. "He's the one who gave me the push from MonCul."

The man who received us was indeed the Dr Sneed who had been employed by MonCul.

"Welcome to Panama," he said. "I can't say it's nice to see you again Dr Frank. Would you introduce the rest of your companions?"

Frank did as he was told.

"Now," said Sneed, "why are you snooping in Panama? And don't tell me you are on a vacation cruise. The lollipop found in your possession betrayed your real purpose. You have come to destroy my career. As for you, the Spandau twins, I know you were responsible for downsizing MonCul, leading to my dismissal."

"There was nothing personal," said Orville. "You were not qualified for pharmaceutical research."

"Not qualified!" screamed Sneed. "Not

qualified? Who was it that joined McDuff's and perfected TMSG."

"Which is now destroying large numbers of Fiddler Crabs and Californians," said Angela.

"I remember you," said Sneed. "You were the insolent young pup who insulted Dr Trickie at a MonCul meeting. You are displaying your ignorance again. TMSG saved the world's panda population. When the last bamboo forest disappeared, the starving pandas were fed French fries, laced with TMSG. Their population has survived and even increased."

"Come off it, Sneed," said Frank. "You may have saved the pandas, but you are now in the lollipop business and probably making billions out of the addicts you have created."

"That's a lie," said Sneed. "TMSG is not addictive. I have been sucking ten lollipops a day for six months, and have not been affected.

And we have put a warning on the wrapper advising parents not to give them to children under three."

"Admitting the lollipops are addictive," said Mary.

"No way!" shouted Sneed, feverishly stripping the wrapper from a lollipop. "The warning is there to prevent dear little children from choking." He began to suck on his lollipop energetically. "I am satisfied that you are a danger to the Panamanian economy. I therefore sentence you to perpetual seclusion in a place where you can't get up to any mischief."

Chapter Thirteen

General von Rouse leaned back in his chair and placed his feet on a human footrest.

"Keep still," he rasped. "I'm thinking."

Rouse was thinking about his bid to take over as Chairman of the WO, the pinnacle of world power. When he left Brazil, he had a price on his head, but that didn't matter because he had 180 billion in the SeaScam cruising bank. Central Snoopers warned him that CoCoCo, the successor to MonCul, intended to impound the ship as soon as it docked in a port. They would have to get up early if they thought they could put one over on him. After brief negotiations, he had bought Panama from the incumbent dictator, Carmen Joinus, for 100 billion and brought the SeaScam into port, now his port, where it was safe from being impounded. The unfortunate Carmen Joinus had not lived to collect his money, since he suffered a fatal heart attack while on board the SeaScam.

Rouse delved into his in-tray and read a message

from Gerbil Sneed, informing him that a party of intruders had been detained and sent into permanent seclusion.

"So what's new?" he muttered. "My camps are full of such people."

"Sir?" said the footrest.

"Shut up and keep still," said Rouse, digging his spurs into the footrest.

The General often visited the camps. He would stroll through the groups of detainees, slapping his shiny jackboot with his riding crop. He would occasionally stop and give an inmate a lollipop, note the number tattooed on the prisoner's forearm, and tell a guard to open a medical file on the lucky recipient. He was in his element. Things had certainly looked up. He no longer felt guilty about the deaths of Trickie and Rupert Bach-Hand. He had raised a memorial to them in the grounds of MonCul, stating that they had died on active service. He had also paid for a memorial feast at the Universal Vegetarian Church and Restaurant. The shades of Trickie and Rupert had been laid to rest.

He was just about to leave his office to go on a camp inspection when a short, swarthy man in shades walked in and sat down. Luigi Parmesano was over eighty, but sprightly. The natives of his Sicilian village believed he had sold his soul to the devil.

"Hello, Luigi. What's your problem?" said Rouse.

"You have the problem," said Luigi. "My family members in LA and San Francisco tell me you have taken their business. Now you are doing the same in New York."

"Luigi, my brother," said Rouse, "I'm marketing lollipops. Can you blame me if your customers prefer them to your product?"

"I'll tell you what we are going to do," said Luigi, his voice full of menace. "We are going into partnership."

"I don't see why I should," said Rouse.

Luigi took his shades off and fixed Rouse with a stare that made him shudder.

"I agree, Luigi," he said. "We'll combine our distribution networks and share the net profits."

"We share the gross take," said Luigi.

"Fine," said Rouse, with a nervous laugh. "There's plenty of business for both our organisations."

"Why lollipops?" said Luigi. "Why not tablets?"

"Sucking a lollipop," explained Rouse, "allows slow release of the TMSG. The stick facilitates the removal of the lollipop from the mouth, further controlling the release." He opened a drawer and took out a handfull of the latest batch. "Try these, Luigi. Give them to your grand-children."

Luigi was on his feet in a flash, a pistol in his hand. "You trying to poison my family, you lousy Kraut? I heard about you and your medical experiments."

"No, no, no, Luigi," said Rouse in a panic. "Forget I ever mentioned them. You see they are quite harmless and very enjoyable."

"According to some people, so are cigarettes," said Luigi. "That's why half my countrymen die of lung cancer. My lawyers will be coming to see you. Make sure they leave happy."

Rouse was worried. He had upset the Don and that was a fatal thing to do. He thought of having him liquidated, but he would never get away with it. Luigi had a Mafia network in Panama, which controlled PoliceCorp and the

Panamanian Bureau of Central Snoopers. Climbing into his ancient Tiger Tank, which had been present at the Battle of the Bulge, he roared down to the docks. He went on board the SeaScam and into the Bank Manager's Office.

Yodel Lay was signing papers, which triggered the transfer of several billion dollars from an impoverished West African country to his bank.

"Busy, Yodel?" said Rouse.

"Never too busy to see you, General," said Yodel. He pointed to a map on the wall behind him. "If you have any investments in that benighted country, cash them in. It is about to be bankrupted by the President."

"That's the third this week," said Rouse. "Why so many, suddenly?"

"They have heard that SeaScam has found a haven where it can't be impounded," said Yodel. "And your abolition of all extradition treaties means that they will be safe here."

"The fools!" said Rouse. 'Don't they know that I have passed a secret decree allowing me to confiscate the proceeds of crime?"

"Well, no, General," said Yodel. "You haven't actually made any seizures yet."

"That's because it is hard to define the phrase 'proceeds of crime'," said Rouse. "The line between legitimate profits and illegal gains has become blurred. Anyway, the treasury is in a healthy state at the moment. When we need more funds, I'll look at the fugitive presidents."

"Is there something I can do for you, General?" said Yodel.

"There is," said Rouse. "I want you to buy diamonds for me. Do it slowly so as not to push up the price."

"And what figure did you have in mind?" said Yodel.

"The whole of my 180 billion," said Rouse.

"You're not thinking of doing a runner, are you, General?

"Not at the moment, Yodel," said Rouse.

"If you do decide to move on to fresh fields, you will let me know."

"Of course," said Rouse. "What are friends for?"

Rouse drove his Tiger Tank across Panama thinking: they don't make vehicles like this anymore seven million kilometres on the clock and still going strong. He arrived at the Number One Holiday Resort and Detention Centre where newly arrested undesirables were accommodated. His tank roared through the perimeter wall, which had been built by the American Corps of Engineers.

"Fix that hole at once," he screamed at the guards. "And make the wall stronger while you are at it."

He dismounted and strode across the exercise yard, swinging his riding crop, striking his boot and any unfortunate prisoner who got in his way. Suddenly he stopped. I saw he was looking at us. I scuttled inside Mary's blouse just in case Graft remembered me.

"What do we have here?" he said.

Frank was the first to speak. "You have five people who are going to make sure that you spend the rest of your life in jail."

"Brave words, Doctor," said General von Rouse, alias Gabby Bogart, alias Heinrich Graft. "So we meet again. And who are these miserable specimens of inferior humanity?"

"Go to hell, Graft," said Frank.

Rouse swung his riding crop, raising a red welt on Frank's cheek. "I didn't know you were here," he said. "That weasel, Sneed, has some explaining to do. Who are your companions? I recognise the delightful Angela and the impertinent Contrary Mary, but the two men?"

Frank remained silent. I picked up Rouse's brainwaves. He had no idea he had the founders of MonCul in his power, and he was completely insane.

'Guard!" he bellowed, "handcuff these people and bring them to the Presidential Palace. I am about to do some entertaining."

We were bundled into a truck. Rouse mounted his tank and crashed through the perimeter wall, making a second wide breach. We followed on in the truck. In other circumstances, I would have found Panama a pleasant tropical city. I got into silent communication with Coco.

"What gives, old Buddy?" I said.

"I'm still on the boat," he said. "It's tied up in

the port with no-one on board. All I can do is monitor the airwaves."

"Any joy?" I said.

"Well," said Coco, "Luigi Parmesano is really p. o. with Rouse. He has asked the Panamanian Mafia to rub him out, as he put it. They are reluctant, because they are doing very well out of his government, which is making life more secure for them. The only hope we have of getting our friends out is to start a campaign of black propaganda and ignite internecine strife. So far, I have no idea how to do that."

"Leave it to me, old Buddy," I said, "When it comes to strategic planning, I have all the right moves."

We arrived at the Presidential Palace and were allocated our suites. Wilbur and Mary were given the Empire Room. It had furnishings, which were mainly made up of the letter N with a few letter Js thrown in. When we were alone, I scrambled out of Mary's blouse and scuttled to a perch on a beautiful Empire desk, on which Napoleon had written letters to the crowned heads of Europe, who were mostly his relatives. Talk about nepotism: Napoleon invented it.

"Thank goodness you are out, Radish," said Mary. "Why did you insist on burrowing inside my bra? I found it very uncomfortable."

"Sorry," I said. "I didn't want to slip further down. I think you would have found that even more uncomfortable."

"Right," said Wilbur. "What's on the grapevine, Radish?"

"There's a lot of gossip," I said, "which may be entertaining but which is irrelevant to the tasks in hand, to wit: primo – how do we escape from this dump; secundo – how do we stop

Graft/Bogart/Rouse from taking the top job at the WO?"

"To take your secundo, primo," said Wilbur, "The current incumbent at the WO is Ginseng Root, because, I suppose, it was China's turn for the post. How does Rouse propose to dislodge a root?"

"He's busy bribing the Electoral College with huge sums from the Panamanian treasury and giving them Vanuatu passports as an optional extra," I said. "Fifteen of the Electors have already taken the money and the other five have suddenly passed away. Their replacements are all ready to take the money. As soon as they are sworn in, an agent will slip a powerful herbicide into Ginseng Root's bedtime cocoa, and they will select Rouse as the new President of the WO."

The future looked black. We wondered what exquisite torture Rouse had in mind for us. We soon found out. He decided to brainwash us into joining him in his dream of world domination led by a strong leader, himself. We were forced to watch archive film footage of endless maniacal speeches by the worst dictators who had ever blighted human kind. The tirades were in German, Japanese, Italian, Spanish, Russian, Portuguese, Arabic, English and most of the many African languages. The only lesson they imparted was to remind us that most countries had suffered under a sadistic ruler at some time in their history.

Coco was quite unmoved. "People," he said, "get the government they deserve."

When Rouse heard us laughing at the strutting megalomaniacs on the screen, he was furious and sent us back to the Number One Holiday

Camp and Detention Centre.

"You people are going to get the accommodation you deserve," he screamed. "Get out of my sight!"

Chapter Fourteen

D'Arcy Versey, wearing an immaculate white suit and Panama hat, descended the gangway from the Ocean Princess onto the dock. A single porter came after him carrying a D'Arcy designer label suitcase. D'Arcy was travelling light again. As usual, he carried a briefcase containing his rainy day wads of hundred dollar notes. His visit to the States had been a success and he was returning to his devoted Honey Bunny with lots of money and a rabbit skin to wrap Baby Bunting in. Yes, D'Arcy was the proud husband of Conchita and father of baby

Astair, Fred for short. D'Arcy had a diamond ring for Conchita. He had never forgiven himself for allowing that dragon, Gladys Spandau, to confiscate the one he had bought for Conchita on Gladys's credit card.

D'Arcy had been in Miami to drum up some business. He had been a roaring success with the blue-rinse brigade. His Tango School had been the talk of the town and the ladies had fought to take lessons with him. He was a brilliant dancer and a great teacher. However, since nobody is perfect, he had persuaded five widows that he intended to lead them with a romantic twirl to the altar. The five had contributed large sums to pay for a cottage with roses round the door. D'Arcy had the cottage money in his briefcase. He intended to apply a small portion of it to his favourite charity: the D'Arcy Versey Arts

Foundation. Although he was a conman, he had his good points.

The dancer had served three years in an Arizona low security prison. Many of the inmates were people like himself who had broken a few financial rules. Many of them were interested in the arts, and not a few in the theatre. The musical theatre had been dead for many years, but D'Arcy had devoted his energies to reviving the great 20th century musical classics. Having found a captive audience that was appreciative of his work, he continued it in Panama.

The Number One Holiday Resort and Detention Centre was fertile ground for him. He got hold of the list of inmates and decided there was sufficient talent to put a musical on. There was a Hungarian Gypsy Orchestra, a group of

Mongolian dwarf acrobats (or maybe they were small children), a British Rock Group, and a touring production of Romeo and Juliet. They had all been incarcerated by General Rouse, because they were either not racially pleasing or stuck-up Brits. It was this wealth of talent that drew the dancer to visit the camp. He immediately spotted a group of people he knew.

"Hello," he said. "Fancy meeting you here."

The twins were surprised to see their mother's ex-pal but were polite enough to respond.

"Hello, D'Arcy," said Orville. "Are you inside or just visiting?"

"Just visiting," he said. "I have General Rouse's permission to stage a musical and I'm here to cast it. He wanted a Wagner extravaganza, but I persuaded him that it would be too much for amateur talent and anyway it wasn't a musical. We are going to do *The Sound of Music*. It will mainly be dance numbers with the actors miming the songs. The Gypsy Orchestra will be there to add a little verisimilitude and play the overture and intermission. The Shakespearians will take the speaking roles, of course. The audience will love their Limey accents. I do hope you boys have forgiven me. Your dear mother has. I met her in Miami and she didn't blow the whistle on me, so she must have."

"I know *The Sound of Music*," said Mary. "We did it as a school production. The family escapes the Nazis. Do you think we could emulate them, D'Arcy?"

He looked at Mary. "Maria!" he said. "You've gone all white. What happened?"

"I stopped taking the pills," said Mary. "Do you think our escape can be arranged?"

"I don't see why not," said D'Arcy. "With everybody hypnotised by the final, magnificent production number, you could slip away in the properties truck."

The dancer set about casting the production. He sorted out actors: Angela was just right for Maria; Mary, the Mother Superior; Orville, the captain and Wilbur, his friend. The Chinese dwarf acrobats were to be the von Trapp children. He gave the Gypsy Orchestra the music and told them to practice playing it at half their normal speed. The Rock Band was asked to be Nazi Storm Troopers (talk about typecasting). All the inmates interested in dancing were rapidly formed into ranks of eager hoofers. Frank was unable to remember lines, so he was given a non-speaking part as a nun. He always seemed to finish up as a female impersonator.

I was in constant touch with Coco. I told him

about the planned Great Escape. He immediately phoned Boats-R-Us and hired a captain to get the boat ready to put out to sea. It took a while for Captain Jaime to get used to taking orders from a nut, but he had served under some funny owners in his time. He provisioned the boat for the trip back to LA and topped up the fuel tanks. When a PoliceCorp man asked him why he was working on the boat, he said he had been hired by a rich nut. Thinking that General von Rouse was the rich nut referred to, the PoliceCorp man moved on.

The General was in urgent communication with his contacts at the WO every day. He felt he had to move quickly to avoid being liquidated by the Parmesano family members, who were furious at losing out to the lollipop craze. The deal that Luigi's lawyers cut with Rouse was only a stopgap measure. No self-respecting gangster could agree to such a humiliating loss of face. Yodel Lay was buying up diamonds as fast as he could and was unable to prevent the price from rising. General von Rouse had a tailor fashion a combat jacket of many pockets, not for ammunition, but to accommodate diamonds. Coco and I were aware of the General's preparations and kept our friends informed. Unable to do anything about the matter, they concentrated on rehearsing for the show and the Great Escape.

The night of the show came round and the cast gave a brilliant performance. The can-can, performed by the nuns in the second act, brought the house down, and D'Arcy's tango with the Mother Superior received a standing ovation and an encore. The Hungarian Gypsy Band infuriated Rouse by playing patriotic folk tunes during the intermission. He made a note to include them in his experimental programme. The Chinese dwarf acrobats charmed the audience by doing back flips and double somersaults while singing the Doh-a-Deer song. When the final production number went into a glorious climax, our party slipped away. During the final scene, Burp Loudly, the Rock musician, who was playing a Storm Trooper, ran onto the stage and shouted, "They've gone! The von Trapps have gone."

Thinking it was part of the show the audience applauded.

"No, you idiots!" screamed Burp, "They really have gone, and I'm following them."

With that, he ran for the gate. For a while, the inmates were not sure how to react. Suddenly someone shouted, "Let's go!" and they rose en masse to scramble for the exit and freedom.

We were crossing Panama City in the properties truck. The Chinese dwarf acrobats came with us, because Wilbur refused to abandon them after they had played the von Trapp children so charmingly. He and Mary were childless and were looking forward to adopting them.

"Now you can do a remake of *Snow White*," said Orville, who was wondering how we would all fit into the boat.

We arrived at the jetty but we didn't have a problem with fitting into the boat, because a squad from Central Snoopers was waiting for us. We were hustled unceremoniously, not back to the Number One Holiday Resort and Detention Centre, but on board the SeaScam, which was getting ready to leave port. We were locked into second-class cabins. Coco and I were with

Wilbur and Mary and their seven adopted dwarfs. It was a tight fit.

"So much for the Great Escape," said Mary, huddling up with her children, who were already calling her Mama. "Thinking about it, our boat was the obvious place for us to run to. We should have had more imagination."

Coco and I considered post-mortems a waste of time, and were busy monitoring the traffic between the SeaScam and the General. He had just missed being assassinated, when a bomb exploded under his Mercedes. He had had the foresight to tell his double to ride in the car while he travelled in a hearse. This meant that observers crossed themselves as he passed instead of blowing him up. In his stateroom on the ship, he was filling his combat jacket pockets with diamonds, and talking to Yodel Lay.

"We leave at midnight," he said. "Guido

Parmesano is already ensconced in the presidential Palace, eating my Patagonian Tooth-Fish and drinking my schnapps. His next move will be against our bank."

The captain was already setting a course for the Mediterranean, with the ultimate destination, Naples.

"See Naples and die," said Wilbur when I told him. 'Why is Rouse taking us there?"

"He wants you out of circulation," I said. "He thinks you might interfere with his plans to take over the WO if you are free."

The ship was going to Naples, but Rouse was going on to Rome. The WO was housed in an area of the city once called the Vatican. Many of the ancient customs of the Eternal City had been adapted to suit the world body. The Delegates from the Multis were called Cardinals. They chose the President and announced his election with a puff of white smoke and the cry 'Habemus Ducem'.

Throughout our stay in the Number One Holiday Resort and Detention Centre, I had tried to keep in touch with the UCLA authorities. Unfortunately it was vacation time and only one janitor was on the campus and he didn't believe a word I said. I kept trying now we were on the SeaScam with the same result. I contacted PoliceCorp; they threatened to arrest me for wasting police time. Central Snoopers was in thrall to the Parmesano family. It looked hopeless. We settled down to wait. Wilbur and Mary spent what they called 'quality time' with their brood. (I never did understand the phrase.) Frank and Angela helped with the children. Orville started planning our escape using materials salvaged from the trashcans. He had already made each of us a splendid life jacket from plastic tablecloths and polystyrene chips. The days passed, the indigo sea rolled by, gradually changing to blue as we entered the Mediterranean.

Chapter Fifteen

Luigi Parmesano was in his Sicilian villa talking to Dr Richard Trickie. Yes, indeed, it wasn't a ghost, it was the cross-eyed scientist in person. (How's that for a *deus ex machina*?) Trickie was not murdered in Tangier. He had agreed with Luigi to disappear and join his organisation as an R and D man. It was either that or a pair of cement shoes. In Tangier, a corpse, identified as Trickie from the passport found on the body, was buried in the European Cemetery. For a small consideration, PoliceCorp supplied the necessary papers and reported the tragic murder. Trickie returned, incognito, to the secret Parmesano Pharmaceutical Laboratory somewhere in the Sicilian hinterland.

Luigi was explaining to Trickie that he had chosen him to run for President of the WO.

"What's wrong with Ginseng Root?" said Trickie. "He's doing a good job."

"That Root is a big disappointment," said Luigi. "He refused to sign release papers for my nephew, Sacco. He has to go."

"But Sacco killed three judges before he was caught," said Trickie, immediately realising he had said the wrong thing. "He did good work for the family."

Luigi put his pistol away. "The SeaScam has docked in Naples," he said, "and Rouse is on his way to Rome. He thinks he is going to be the next President of the WO, the moron."

"Isn't he?" said Trickie. "I thought it was all arranged for him to take over."

"So how is it that I am going to put you in the job?" said Luigi.

"I really don't know," said Trickie.

Coco and I knew. For weeks, we had been intercepting messages to all the important Mafia families, inviting them to a meeting in Palermo. It was held in Luigi's villa and I made sure that EvesDropSat was positioned overhead to relay the proceedings directly to me.

Luigi addressed the assembled Dons. "Brothers," he said, "Ginseng Root is a useless tuber and he's getting dangerous. Sacco is still in jail so we have to do something about it. The WO Cardinals are from the Multis and we do not have a single voice in their deliberations. It's time for a change. I propose to put our representatives in, with a Parmesan Cheese as the President."

Dr Trickie, who was taking the minutes, smiled. He had been replaced at the last moment by a cheese. A cheese at the head of the WO, he thought, what a joke! There was a stirring of dissent in the assembled Dons. Asiago Pressato from the Po Valley spoke first.

"My respects, Don Parmesano, but why not vote for the President? We have here present, Don Baita Friuli from Venice, Don Sottobosco from Alba, Don Brillo from Treviso, Don Bosina Robiola from Piedmont, need I go on? Any of these cheeses would make a good President."

"You have a point," said Luigi. "We'll take a vote at the end of the meeting."

The vote was inconclusive, so the matter was left open. After the meeting, Luigi said to

Trickie,

"I decided not to put you at the head of the WO. I knew it was the right decision when I saw you smiling at the idea of a Parmesan cheese in the position. Anyone who can make fun of a cheese is not fit for high office. I'll let the insult go this time. The next meeting of the WO Cardinals is on St Valentine's Day. We have arrangements to make."

When February the 14th came round, the WO Cardinals were in the Vatican Cellars. Swiss Guards, captained by Andre Gide, with machine pistols cocked, stood guard over them. Luigi and his fellow Dons came in and handed out two sheets of paper to each delegate.

"You have a choice," he said. "Sign these and you will be allowed to go back to your countries and run your Multis. The first one is your letter of resignation from the board of the WO. The second one approves the list of your replacements."

"I for one am not signing anything!" shouted the CEO of the World Football Association, a particularly stupid man. He was dragged out, placed against a wall and torn to tatters by a burst from a machine pistol.

"Pour encourager les autres," said Luigi, quoting what Voltaire said about the hanging of Admiral Byng for refusing a sea battle. There was a great scratching of pens on paper. Luigi addressed the Cardinals, "Where's General von Rouse?"

They didn't know. The General, a careful man, had intended to arrive after the preliminaries, which were due to last an hour. When his scouts saw Luigi and the Dons turn up with a regiment of Swiss Guards toting machine pistols instead of halberds, they smelled a rat.

Luigi was furious. "I'll deal with the General later," he said. "My brothers, let us go up into the Council Chamber and elect our Chairman."

Ginseng Root was chopped into little pieces, infused and served to the delegates. In recognition of Luigi's work in taking over the WO, the Dons elected a Parmesan Cheese to the top job. Then everybody wondered what to do next. The fact was that the WO's power was in the hands of the Cardinals who were all CEOs of huge Multis. The only influence the Dons had was over vineyards, olive groves, flocks of sheep and goats and water supplies. In the following weeks, the Multis simply ignored the WO and ran the world using electronic conferencing facilities. Coco and I were able to follow the new style of world government and witness the demise of the Mafia controlled WO.

The Dons stopped attending meetings at which nothing was done. They tried declaring war on Monaco but the principality simply issued the famous declaration, *Faites vos jeux!* and ignored the threat. The WO was dead. Their premises, in the former Vatican, were taken over by The Church of Seven Day Vegetarians, which had become the world's dominant religion. Richard Trickie went back to his lab and started work on finding the Philosopher's Stone. The only change he wrought with his Black Magic was to transform himself from a reasonably normal crook into a madman.

On our arrival in Naples, we had been detained on the SeaScam for a week to allow the General to take up his position at the WO. When this did not materialise, we were let go by Yodel Lay with a warning not to talk to the media. We flew back to Panama where we all went to see D'Arcy Versey in the *The Sound of Music*, which had transferred from the Number One Holiday Camp and Detention Centre to the Panama Opera House. The show was attracting record crowds and giving employment to the Hungarian Gypsy Orchestra and many other artists released from the detention centres, which had all been closed. We went backstage during the intermission to congratulate D'Arcy.

"Things are looking up," he said. "I have had an offer to take the show to New York, and there is a suggestion that we might do a re-make of the movie with me in the Christopher Plummer role. Would you consider investing in my venture?"

Orville and Wilbur politely declined the generous offer. D'Arcy had not changed. Frank got in touch with the UCLA office and was told there was no such person as Dr Francis Frank teaching there. Due to his unexplained absence, a new man had been appointed and Frank's name expunged from the records. We reclaimed Frank's boat, which was being used for fishing trips by Jaime who hated to see a good boat permanently in dock. Frank thanked him for maintaining the boat and paid him to re-provision it and top up the fuel tanks for a second time.

Wilbur and Mary were delighted when they found out that their charges were not, in fact, dwarfs, but children aged from six to eight. This resulted from taking a crash course in Mongolian so that they could talk to them. They decided to fly to Florida to see the children's Grandma, Gladys Spandau, and take a side trip to McDuff's World, the huge theme park. The manager of the park recognised the children as

The Delightful Dim Sums, one of many troupes of acrobats produced by the Mongolian State Orphanage.

We paid a farewell visit to Crab and her two babies; they were much happier now that the spilled lollipops had been exhausted. The Fiddlers were having counselling for their addiction from a Hairy Crab, who was helping them through their withdrawal symptoms

Frank, Angela, Orville, Coco and I sailed back to LA in comfort with Jaime, who expressed a desire to emigrate illegally, captaining the boat. Frank and Angela paid a visit to the UCLA campus where they were unable to trace their stuff, which had been left in their campus bungalow. They turned down an offer to join Orville and Wilbur's large family in Culver.

"It's a nice enough place," said Angela, "but we

don't want to live so far from the ocean. Have you ever eaten seafood in Culver? Yuk!"

Coco decided that Q was doing a great job as Chairman of CoCoCo, and that the cucumber should continue in that position. He stayed with Frank and Mary and took up the study of Zen Buddhism, which was having a revival. I had become very fond of Wilbur's new family, so I decided to live with them. The seven children and I worked out a routine, which included me doing a triple somersault, from a great height, and landing in a damp cloth. People watching screamed in horror and then applauded as I bounced up and took a bow. Things were getting back to normal.

Chapter Sixteen

General Rouse was on the bridge of the SeaScam, back in Panama and renamed The

Bismark. He watched as his Tiger Tank was hoisted on board and thought about his last moments in Rome. After the coup by the Mafia, he had assumed the guise of an Algerian carpet seller. He was most surprised when he started making large profits from carpet sales. He watched as the Dons assumed ultimate power, let it slip like sand through their fingers, and slink back to their mountain lairs. He then had the galling vision of The Church of Seven Day Vegetarians moving into the WO buildings. He pondered on the rise of the SDV church and wondered why he had never had such a brilliant idea. The founder, Sam Small, a small farmer, had been working on his smallholding, when he heard a small cabbage speak to him in a small voice.

"Read my leaves," it said. "On them you will find writ the eternal truths of our faith."

Sam read what was writ and even writ the words in his farm record book, which can still be seen at the headquarters of the church in Washington. The first leaf started with, 'And in the beginning the earth was more like a cabbage.' It was a message, one that Sam propagated throughout the entire world. The rest is history. Unfortunately, cabbage leaves are not a good medium for recording the written word, so they perished and were lost. One schismatic sect maintains that a second cabbage will come soon to reveal further truths, but nobody is holding their breath.

Panama was a dangerous place for General von Rouse and Yodel Lay. Guido Parmesano was in the President's chair, busy setting up casinos and money-laundering banks. To avoid competing with the Mafia businesses, production of the detested TMSG lollipops had ceased in Panama, but was starting up in Colombia. Gerbil Sneed had gone underground for a while and then turned up at the Bismark seeking asylum. The General welcomed him aboard.

Guido didn't realise that the Bismark was the former SeaScam or he would have come down on the vessel like a ton of bricks. This gave the General time to fit two batteries of six-inch naval guns fore and aft on his ship. His submarine, which had carried lollipops, was renamed U66, fitted with torpedo tubes and a deck gun, and manned by an experienced crew, average age eighty. Before leaving Panama, the General had Yodel Lay purchase an uninhabited island in the South China Sea. His intention was to declare the island an independent monarchy, where the Bismark would be safe from sequestration. As the fleet sailed out of Panama harbour, the General looked back and said the immortal words, "I shall return."

The voyage to the South China Sea was uneventful. To give his gunners practice, the General had them sink a fishing boat, and his submarine torpedoed a cruise liner. He thought about the great days of the 20th century when such activities were common and wished he had been born then. The island they were heading for was in the Diaoyu Group. Yodel considered it a bargain at a billion dollars. When the GPS indicated that they were approaching the island's co-ordinates, the General scanned the horizon with his binoculars. He saw that the island was barely a foot above sea level. As he watched, the tide came in and the island disappeared. He stormed down to Yodel's office.

"You crook!" he screamed. "How much did the real estate agent pay you to buy that lemon?"

"I only took my usual 20 % commission," said Yodel.

Rouse was about to shoot him when a tremendous explosion shook the Bismark. Scampering to the bridge, he saw the fleets of China, Indonesia, Japan, Korea and the Philippines engaging in a fierce sea battle. Waves of suicide bombers were taking off from the decks of the carriers. Surface-to-surface and surface-to-air missiles were whizzing towards their targets. Attack fighters were rocketing and high altitude bombers were bombing. The Battle of the Diaoyu Islands turned out to be the biggest since the Battle of the Coral Sea, which, lost in the mists of time, was only recorded in a few history books. The General awarded himself a battlefield promotion to Admiral, and ordered his fleet to sail away from the conflict, sinking anything that got in the way. Sailors from sunken vessels clambered aboard the U66, which had to submerge to shake them off. Some tried to board the Bismark using a ladder left carelessly over the side. A machine gunner persuaded them to leave. It appeared that all the combatant countries had laid claim to the islands, which only surfaced at low tide. They were fighting for control over them, which, to Admiral Rouse was a perfectly good reason to have a battle.

To punish Yodel Lay, the Admiral found a desert island, which was not submerged, and marooned him on it. He was allowed to take a hand-cranked gramophone and eight records plus a choice of two books, in addition to the Vegetarian Testament and the complete works of

President Dubya (a slim volume consisting of blank pages). Yodel returned to Panama two years later with his partner, Friday Knight, who he had met on the island, which was not in fact deserted, but chronically overcrowded with people who owned timeshare huts. Yodel resumed his career as banker and prospered under a laissez faire regime ruled over by a Pineapple.

Admiral Rouse had a present score to settle with a real estate agent and Guido Parmesano. The Bismark steamed into Panama Bay accompanied by U66. As the submarine pumped torpedoes into the harbour, crowded with shipping, the Bismark fired broadside after broadside into the Central Business District and the Presidential Palace. After the Battle of the Diaoyu Islands, the Admiral had obtained fresh torpedoes and shells from Parmesano Industries. Fortunately for the Panamanians, he was sold munitions, which had passed their sell-by date and did not explode. In a blind fury, he took the Bismark into the harbour, hoisted his Tiger Tank onto the jetty and roared across the capital. At the Presidential Palace, Guido Parmesano goose-stepped out into the courtyard, a disdainful sneer on his face.

"You fool," he said. "I sold you the dud munitions. Now you can have the real McCoy free of charge."

He raised his anti-tank bazooka, but before he could pull the trigger, Rouse got off a round, which splattered Guido against the wall of his palace. He had forgotten that Rouse still had tank shells.

"What do you think of them bananas?" said Rouse, mixing up his fruit. "Revenge is mine.

Revenge is sweet. Revenge is a dish best eaten cold. And other cliches."

Killing Guido Parmesano was a big mistake. Luigi did a tour of the Multis and bribed the CEOs liberally. They had no reason to like General von Rouse; in fact, they all thought he was a menace, so they were happy to raise a task force to hunt him down. The cry 'Sink the Bismark!' went up and was repeated round the globe. The ship was tracked down to Lake Geneva and sent to the bottom by a well-placed mini-nuke. How the battleship got to a land-locked lake was a mystery, but one could expect anything of the devilishly cunning General. It was just like him to conceal his ship on a lake, which was impossible to reach. The Bismark at the bottom of Lake Geneva was in fact a disused car ferry, which had plied the lake for many years. It had been camouflaged to look like a warship and given the new name.

When the genuine Bismark sailed back into Panama Bay, it had been transformed into a psychiatric hospital ship, renamed *The Couch*, ostensibly working for Shrinks without Frontiers. This was a wonderful organisation, founded by Sister Agnes, whose ship travelled the oceans calling at countries torn by wars and civil strife. The Sisters of Commiseration were all trained psychiatrists, who went ashore and comforted the populace. Anyone could supply food, shelter and medicines, but they provided something more valuable: sympathy.

Sisters Agnes Rouse and Greta Sneed walked the deck in silent contemplation.

"What's new, Sister Greta?" said Sister Agnes.

"Some pansy dancer is in the Presidential Palace," said Sneed. "Name of D'Arcy Versey."

"He's the clown who organised the Great Escape," said Rouse. "How did he get to be President? Couldn't they find a vegetable?"

"It seems not," said Sneed. "They decided to have an election. Versey, being a star of stage, screen and the penal system, was the best known, so he got elected."

"Don't people know how to rig an election these days?" said Rouse

"Yes," said Sneed. "It was Versey who rigged it, buying votes with tickets for his new musical, The D'Arcy Versey Story."

Chapter Seventeen

I was Wilbur and Mary's Social Secretary, busy taking calls and organising their schedules now that the children were fluent in English and able to mix freely with the children of Culver. I arranged piano lessons, birthday parties and excursions and tutored them in a range of subjects. Angela called to say that Frank had got a job as a Marine Biologist. She explained that he would be the Chief Scientific Officer on the Ocean Explorer, a survey ship, mapping the remaining coral reefs and recording the decline in fish and sea mammal species. They were taking Coco with them and wondered if I would like to go along for at least one trip. Frank came on the phone and explained that I would be an asset in that my GPS, Sonar and other features would be a good backup in case of storm damage to the ship. I promised to talk to Wilbur and get back to them.

Wilbur and Mary were reluctant to let me go. The children were all against the idea. We did a deal with Frank and agreed to pay for food and accommodation for the whole family. The Foundation, which owned the ship, did not object, as our fee would pay most of the expenses of the expedition. Following our example, the Ocean Explorer eventually became a floating school, running educational tours.

We joined the ship in San Diego and set off down the West Coast. There was great excitement when a diver discovered a coral reef that was starting to regenerate. The sad fact was that the rise in sea temperatures had killed off most of the world's corals. It was only because fossil fuels had become scarce and very expensive that the process had slowed down. The very rich were still driving SUVs, but the horse was making a comeback in some countries, much to the benefit of the air quality and people's roses. The children were delighted with the dolphins that joined the ship, leaping ahead of the bow wave. Our young acrobats did cartwheels on deck, encouraging the dolphins to emulate them. Our tour was to take us into the Atlantic, which meant passing through the Panama Canal.

D'Arcy Versey turned out to be a very successful President of Panama. The canal had been lost in a takeout by McDuff's and no longer produced any revenue for the country. In spite of this, D'Arcy followed a regime of bread and circuses, guaranteeing the population unlimited free tortillas and regular entertainment. He restored the musical to its proper place in the Pantheon of the Arts, making Panama the entertainment capital of the Americas. Touring companies were trained and despatched north and south. Revenues from these meant that D'Arcy was able to balance the national budget with a little left over for himself. He put on a gala performance of *West Side Story* for our visit and persuaded the Delightful Dim Sums to put in a cameo appearance as break dancing street urchins.

I began to take an interest in the Shrinks without Frontiers ship, the MV Couch, which was in the harbour. There were many examples of its class, which was produced by a Saudi Arabian shipbuilder. After the oil ran out, the country had put all its resources into heavy industry. I knew the SeaScam was an example of this particular class of nuclear-powered ship. I monitored the electronic traffic and discovered that billions of dollars were being transferred every day, an activity not usually carried on by an NGO. I told Frank and Orville who were of the opinion that it was not illegal for a charitable organisation to run a bank. D'Arcy said he was a patron of Shrinks without Frontiers which was doing good work in Panama, persuading the citizens that virtue was its own reward and loving your neighbour was better than robbing him. He offered to show us round the twenty-couch psychiatric hospital.

Sisters Agnes and Greta watched as we came on board. Their eyes narrowed as they recognised us. SwF was a genuine NGO founded by Sister Agnes. She had considered starting a religious sect because of the tax breaks, but decided that building churches was too costly and she didn't know much about the mystical qualities of vegetables. She selected the Shrinks idea because it was cheap and was badly needed in an increasingly mad world. The advantage was that it allowed her to run an offshore banking operation and laundry without raising suspicion.

Sister Agnes, alias Admiral Rouse, General von Rouse, Gabby Bogart and Heinrich Graft, was still as mad as a hatter, but his mental state had improved since starting therapy with the Mother Superior. She coaxed him into recalling his childhood, youth and early life where the seed of his madness lay. She was a totally understanding and tolerant nun and didn't mind that Heinrich was in the habit of wearing one. If cross-dressing was his bag, it was fine with her and he did finance the whole SwF operation. Her only request was that he should refrain from smoking cigars while walking round the ship, dressed as a nun.

As a boy, Heinrich had been like any other child.

He pulled the wings off insects, tortured cats and inflated frogs with a straw. Once he reached puberty, he cast off childish things and started looking at girls. When he met Helga, he was sure that she was the girl for him. They went to High School and University together, married and set up home in San Francisco, where Heinrich got a job selling kitchen gadgets over the Internet. The company, Penates, realised that their goods were like offerings to the household gods, which were used once and never taken out again because it took too long to clean them. It was a profitable business. Little Hans came along two years later and the happy couple doted on him.

The reason for Heinrich's moral deterioration lay in a traumatic accident when little Hans was six months old. He was sitting in one of the self-propelled shopping carts, which were then in use in supermarkets. They have since been banned because they tended to run over shoppers. Helga was taking a can of beans from a huge stack when the whole pyramid collapsed, burying her. A can hit the shopping cart, putting the accelerator to maximum, causing it to run out of the store and off into the dark, rainy San Francisco night. It and little Hans were never seen again.

Helga was dangerously ill in intensive care. Between visits to the hospital, Heinrich tried fifty-seven ways to trace their baby, going so far as to force the company, whose cans had injured Helga, to put missing person ads on their cans of beans. Nothing worked. When Helga died from the injuries she had sustained, Heinrich cracked. He began to lie, steal and indulge in every corrupt scheme that came along and, to his surprise, he rose rapidly in the business world. He fabricated a CV to suit every new position he sought and finally became the Head of Research at the Interplanetary Travel Corporation. His reputation as a rocket scientist and manager was made. It was this that persuaded Wilbur and Orville to appoint him CEO of MonCul.

D'Arcy Versey introduced us to the Mother Superior, who showed our party round the ship, which had twenty couches for in-patients. Soothing music, mainly Mozart, lulled the patients, staff and visitors alike. The pastel colours of the décor were designed to sooth the savage breast just as much as the music. The cafeteria served a gourmet vegetarian menu designed to eliminate the poisons accrued from eating junk food. SwF was convinced that it was one of the root causes of mental disease. I was very surprised when we were conducted into the Bank without Frontiers. I scrambled to my usual hiding place in Mary's bra.

"This is our benefactor, Sister Agnes," said the MS, "and this is her deputy, Sister Greta."

I peeped out of my hiding place. In spite of their voluminous, black habits and rather fetching wimples, which hid much of their faces, I easily recognised Heinrich Graft and his partner in crime, Dr Gerbil Sneed. They bowed coyly to us.

"The Bank without Frontiers," continued the MS, "provides the necessary finance to allow us to bring sympathetic therapy to the bewildered and befuddled. Our treatment would cost you an arm and a leg in a regular psychiatric practice."

We moved on to the pink, soft-furnished, padded, secure accommodation, which was designed to give violent patients a feeling of being back in the womb.

"Once they come in here," said the MS, "they become totally calm and suck their thumbs. We have difficulty in persuading them to come out."

We took tea in the MS's office. She showed us a video of some of her more difficult cases, which had been successfully treated.

"This is the guerrilla leader, known as the Optician, who chopped off more ears than you and I have had hot dinners," she said. "He had an aversion to people who wore glasses. He went around making sure that fewer people had ears to hang glasses on. Under hypnosis, he revealed that an optician's sign, a huge pair of glasses, had fallen on him when he was a baby. We persuaded him that it was better to remove dangerous signs than ears. As you can see in the video, he now wears glasses for reading."

We saw other cures, including the man who went around stealing babies' bottles. (He had been breast-fed.) There was the woman who fainted when she saw a pile of dirty pots. (She was found to be normal.) There was a man who couldn't stand the sight of children, which caused him to hide in a cupboard. (He was a teacher.) And so on.

As we were leaving the ship, the MS said, "You know, Sister Agnes is a wonderful human being. Don't be surprised if she is recommended for the Nobel Peace Prize and possibly sainthood."

To say the least, we were all nonplussed, flabbergasted even. Could it be that the man who abandoned two donkeys in the desert, who sank a fishing boat and a cruise liner in the South China Sea, who marooned Yodel Lay on a desert island, who reduced Guido Parmesano to pebble-dash, could become a saint? Frank mentioned an example, Paul, who, in the old religion, had proved it was possible.

Back on the Ocean Explorer, we discussed our visit and the prospect of Heinrich Graft winning the Nobel Peace Prize and becoming a saint.

"What is it, exactly, that Graft is doing?" said Mary. She answered her own question. "He's recycling the proceeds of crime and taking a healthy percentage to finance SwF. From what we have seen, the charity is devoted to the task of undoing the damage done to innocent lives by exploitative business interests. I think that is fair and we should not do anything to put a spoke in their wheel."

"I agree," said Coco, who had not been on the Couch but had followed the proceeding through a link I had provided. "I have been reading

Graft's brainwaves. Instead of his normal double lightning flashes, I get a nice, wavy line with no sharp peaks and troughs. In other words, a typical reading for a normal person."

D'Arcy Versey, who had introduced us to the MS, was very quiet. Normally, he would have been monopolising the conversation with accounts of his theatrical triumphs, his future plans and appeals for investors.

"A thought has crossed my mind," he said, "Perhaps all the rascals, I have been fortunate enough to meet in my eventful life, have become boringly conformist, and therefore totally lacking in interest. I propose to leave this vessel and work on some outrageous scheme to bilk some filthy rich parasite of zillions in cold cash or hot dollars. I'll bid you farewell."

Exit one determined conman.

Since we were paying most of the expenses of the Ocean Explorer's foray into the ravaged oceans, we were able to impose our collective will on the ship's crew.

"Right!" said Frank to the Captain, an ancient salt by the name of Jones, "Change of plans, Captain. We're not going through the canal. We're going to the Southern Ocean to find out if there is any marine life left. The Patagonian Tooth-Fish, as I recall, is an endangered species. I want to know if it has regenerated or joined the endless list of extinct species, "

Jones was glassy eyed.

'Captain!" said Frank. "Did you hear what I just said?"

"Aye," he said "I heard ye, ye blubber-eating, blubber of a land-lubber. I'm setting a course for the Southern Ocean to get revenge on the beast that took my left leg."

"But Captain," said Frank, "you have two perfectly good legs."

Jones struck his left leg with a marlinspike and winced.

"Solid whalebone, that is," he said, "carved by an Eskimo craftsman. And I'll harpoon any man who contradicts me."

Heeding D'Arcy's warning about surrounding oneself with conformists, Frank humoured the Captain.

"Aye, aye, Sir," he said, giving a smart salute. "Set all sail and head due south."

"It's easy to see," said Jones, "that you don't know the first thing about ships. This vessel doesn't have sails. It has a mini-nuclear engine."

Chapter Eighteen

As the ship moved south along the Colombian coast, Coco and I tuned in to the signals traffic.

The Colombians, who had been fighting for so long that they had forgotten why they were fighting, were having a conference to find a good reason to continue. Twenty-two factions were represented at the meeting. Some were complaining that the others were ganging up on them from time to time to achieve a temporary advantage. They regarded this as unfair and wanted a ban on amalgamations or alliances between different factions. The main problem with having twenty-two factions was that it was very difficult to decide on who to fight. This dilemma was resolved when two captains were appointed and each took turns to pick a faction. It was the same procedure they had used for picking football teams when they were children. There was only one standout. The First Medellin Boy Scouts (Vegetarian Church Tendency) refused to join anyone else because they were always forced to pitch the tents, tie knots, start fires with two sticks and bake tortillas in a biscuit-tin oven. The Chairman solved the problem by telling their mothers what they were up to, resulting to them being seized by an ear and dragged home.

This left an even bigger problem. Twenty-one factions could not be divided equally. It took several days of argument before it was agreed that one faction, the Motorised Wheelchair Cavalry (Upandatem Tendency) should alternate between the two sides every six months. The paraplegics protested, but agreed when the Sabotage Unit (Sneaky Tendency) threatened to let their tyres down. Choosing names for the two new groupings was fraught with difficulty. The factions ranged from Nuns For Freedom (Ban the Wimple Tendency) to Psychopaths-R-Us

(Obnoxious Tendency). The delegates on both sides wanted the words Democratic, Patriotic and Freedom in their name. Nobody wanted Marxist, Leninist, or Maoist, which were considered old-fashioned. In the end, they agreed to have the same name for both sides: The Democratic Patriotic Freedom Movement, differentiated by the addition of Red Tendency and Blue Tendency. It was also agreed that the combatants should wear football shirts in the colour of their tendency. These would be useful both when they were fighting and when they played an annual football match.

Frank made sure the Ocean Explorer stayed outside Colombian territorial waters. I left communications to Coco and used my equipment to scan the ocean depths for any signs of life. The fishing rights in that area had been sold to a Japanese company that had swept up practically all the fish in one season. An occasional lonely sardine appeared on my Sonar. The dearth of fish meant that everything above them in the food chain had dispersed to other oceans to look for food. Where were the whales, sharks, and dolphins of yesteryear? Gone with the seabirds. We continued south. The Peru Current bringing up colder water from the Antarctic was rich in plankton and able to support a larger fish population. Coco started flooding the airwaves with reports in Japanese, Chinese and Russian that fish stocks were completely played out from Colombia to Antarctica, and the factory trawlers might as well go home and process vegetarian pet food. Many of them did. Frank was elated when it became apparent that the ecological balance was not beyond recovery. I had a chat with an old whale, who remembered Herman Melville sailing those waters. I doubted that he was telling me the truth; I believe he was just relating a story that had been passed down the generations.

Captain Jones became glassy eyed when my friend, Whale, broke the surface to breathe. He slipped into his Captain Ahab personality. I made a note to consult Shrinks without Frontiers to get some advice.

"Thar she blows!" he shouted. "That damned Moby Dick as took my leg. I'll get the better of ye yet, ye plankton glugging mountain of blubber."

He then slipped back into his usual self, remarking, "It's nice to see a whale in these latitudes. It's a sign that things are getting back to normal." He rounded on a deckhand who was staring at him. "What are you staring at, Mr Christian? Swab the deck, ye useless landlubber."

It was becoming difficult to separate Jones from Ahab. As we sailed down the Chilean coast, he said, "It must be strange, living in such a long, narrow country. I imagine it to be quite claustrophobic having the ocean and the land frontier so close together."

The seven Delightful Dim Sums were a constant worry. The ocean was getting progressively rougher and the boat was rolling a good deal. Wilbur, Mary and Angela kept the children in the saloon as much as possible, but it was impossible to restrain them when they suddenly took into their heads to do a series of back flips along the deck. In the end, all acrobatics were forbidden and Wilbur decided to take his family ashore at Santiago before flying back to Culver. I was to be allowed to continue with the ship.

In the Chilean capital, there was a whiff of revolution in the air. Central Snoopers had attempted to overthrow the government and install a military junta. This rang a bell which sent the citizens scurrying to the history books to find a parallel, which was almost lost in the mists of time. It was soon confirmed that such a takeover had happened before with horrendous consequences. So far, the citizenry had held out against the coup. We docked and went ashore to conduct Wilbur and his family to the airport. We passed through several roadblocks manned by soldiers bristling with weapons, machine pistol cocked and backup pistol in the belt. Frank was almost arrested for asking why they needed so many weapons to check identities. It soon became obvious that a small gratuity was in order to ease our path. Coco was in Angela's bag and at one checkpoint a young recruit confiscated him as a suspicious object, possibly a bomb. He flipped open his eye and revealed his teeth.

"Look here, young man," he said. "I am a harmless communications relay station. If you are looking for trouble, I can contact General Spinoza, your commander-in-chief. He will tell you that you can't go around confiscating innocent nuts. Now put me back in Mrs Frank's bag."

The young soldier couldn't obey quickly enough.

We managed to get back to our boat, without getting arrested or accidentally shot. Coco was engrossed in listening to transmissions coming from Central Snoopers.

"General Quagmire, the head of Central Snoopers, has ordered the Air Force to bomb the Presidential Palace," he said. "And Spinoza is leading a column of tanks to take over the government."

Central Snoopers was a huge organization, employing a hundred thousand agents to spy on people and a similar number on spy on them. The General's policy was not to give anyone the benefit of the doubt. His favourite expression was, "Don't even trust your own shadow." On a sunny day, he made a point of always watching his shadow, going to the extreme of walking backwards when the sun was in front of him. His headquarters was the size of a small city. His office was only slightly smaller than a tennis court. His assistant, a nine-foot basketball player, Slam Duncan, had been chosen because anyone smaller would have looked insignificant in the huge office.

"Let me have tall men about me," Quagmire used to say. "Yonder ankle-biter has a mean and hungry look."

Duncan placed a sheet of paper on the General's desk. "Looks bad, General," he drawled. "Spinoza was in too big a hurry to get into the presidential palace and was killed by one of our bombs."

"Too bad," said the General. "But it could be a sign. I never really trusted Spinoza. He wasn't one of us, you know, Duncan."

"Was he one of them?" said Duncan.

"Possibly," said Quagmire. "This will allow us to put a reliable agent into Patagonia."

"Chile, General," Duncan corrected.

"Yes, it is, rather," said Quagmire. "Yesterday, I was talking to Reverend Stir Fry, the Head Chef of the Seven Day Vegetarian Church. He has a Savoy Cabbage that is running their African Franchise. Savoy is a Harvard graduate with a Voodoo qualification. He can kill just by sticking a pin in a carrot doll: a useful skill in a place like Begonia."

"Chile, General."

"Turn the heating up. I want you to organize a neutron bomb strike on the capital. That will clear out all the unreliable locals and allow us in send in an administration provided by GovCorp. Find out if they have got one available."

"I have their latest catalogue here, General," said Duncan. "They have a good range of off-the-peg governments, ranging from stupidly democratic to viciously fascist. What's your fancy?"

"With Savoy Cabbage at the helm," said Quagmire, "we should have a tough but fair administration."

"What about Fundamentalist Vegetarian?"

"Too extreme," said Quagmire. "They go round blowing up butcher's shops."

"The This-Is-Going-To-Hurt-Me-More-Than-You Disciplinarians?"

"Excellent choice for a country like olivia:Terror with a human face: the best of both worlds."

Coco looked round the saloon, where Frank, Angela, Orville and I were listening to one of the leaders of the world's most powerful country with a look of disbelief on their faces, except for me: my leaves were standing on end.

"We should do something," said Angela. "The Disciplinarians kill, torture and plunder. In alphabetical order, they will arrest anarchists, butchers, cattlemen, deaf engineers, fanatics, gurus, handicapped insane jokers, kids, lame mute noisy obese pets, quislings, reds, stupid unbalanced virtual ex-prisoners, yobs and zeros."

"Leave it to me and Coco," I said. "I'll inform the people that Spinoza attempted a coup. We can swamp television and radio with news of what Central Snoopers plan to do, bring them out on the streets and let them elect their own government in a democratic election."

We got to work. Within an hour, the capital was paralysed by crowds of demonstrators. A provisional government with Spinoza's son in charge, was put in place. When Frank questioned the people's choice, I explained that it was the local custom for the heir of an assassinated dictator to succeed him.

Satisfied that we had struck a blow for democracy, we sailed out of Santiago and resumed our research. I had several conversations with blue whales, who were most knowledgeable about the state of the oceans. They were fortunate in that they were able to follow the plankton, but they were worried about their penguin friends, who couldn't follow the fish. Their message was reinforced by an incident, which happened a week later. I was sitting on top of the wheelhouse when I spotted a penguin on a life raft. I told Captain Jones and he lowered a boat to rescue her. We lifted her on board.

"Thanks," croaked Penguin. "Could you spare a small fish?"

Oceanic affairs had sunk to a new low, when a penguin was reduced to begging. The cook produced a bucket of sprats. When Penguin had eaten her fill, she explained.

"The ice shelf where I lived with a million other birds has melted. The water is so warm that the fish have disappeared. We all had to emigrate to find food. I got separated from my family and here I am." A tear trickled down her cheek.

Coco and I did our best to comfort her. Suddenly Coco stopped making funny faces and listened intently.

"Oh dear," he said. "Bad news from Santiago: Central Snoopers has neutron bombed the Presidential Palace; a Savoy Cabbage has been installed as President and a Disciplinarian administration is going in by helicopter. Marines are sweeping through the capital, arresting people in alphabetical order."

The dinner table that evening was quieter than usual. Coco and I sat with Penguin, feeding her sprats, while our friends ate in silence. Orville was the first to speak.

"Well," he sighed, "there's nothing we can do about the situation in Chile. It is my experience, that these imposed government don't last for long. Let's hope the next one is better."

"I have never heard such woolly thinking," I said. "The average life of those administrations is seventeen years. They traumatise the population and when they do go, they pardon all the abusers of human rights. Anything we can do to evict the Disciplinarians will save lives. Coco and I will continue the fight."

I went on deck to cool down. Within seconds I was sending subliminal, black propaganda to the television screens of the benighted nation. Without the authorities being aware, the people were being informed, at a subconscious level, about all the abuses that were taking place. My next move was to persuade a Cabbage White Butterfly to lay a few thousand eggs on President Savoy Cabbage. Within a week, he was dead, stripped to a skeleton by a mass of green caterpillars. With nobody at the helm, the Disciplinarians didn't know who to torture and kill or where to pillage. General Quagmire tried to persuade a cauliflower to take over, but she refused on the grounds that the place was insecure. Our next move was to contact Pancho Zapata, the most respected bandit in the country. We gave him a blueprint dividing the country into a thousand autonomous districts. If Central Snoopers wanted to take over the country again, it would have to kill almost everyone. We went on national television and urged the people to revolt. The plan worked and the new districts were set up. GovCorp withdrew the Disciplinarians, who fled in a fleet of helicopters. Coco and I were cock-a-hoop. We had engineered a democratic revolution. Orville advised us not to get too cock-a-hoop, because revolutions had a nasty habit of deteriorating into totalitarian regimes. Within six months, Orville was proved right: the Seven Day Vegetarian Church had flooded the country with Born-again Turnips and small arms. Central Snoopers had set up a government in the capital, which took control of the country. What can I say? We did our best and it wasn't good enough. And, as you know, it never is.

We continued down the coast of Chile and rounded Cape Horn. Our next port of call was to be Buenos Aires.

Chapter Nineteen

D'Arcy Versey was depressed. His latest production of *Evita* had bombed on a tour of Argentina and he couldn't understand why. The truth was, far from being a folk heroine, the dictator's wife had been completely forgotten. The few people, who turned up at the performances, found the idea of a poor girl, who was no better than she should be, becoming the virtual ruler of the country, rather ridiculous. The result was that D'Arcy was submerged in debt, his passport had been confiscated and he had been forbidden to leave the country. In normal times, he could have drawn on his bank in Panama, but these were not normal times. Central Snoopers had become alarmed because the Panamanians were casting covetous eyes on the canal, which was owned by McDuff's. It was essential for their operations, ensuring that beef supplies could get from Argentina to the west coast. During D'Arcy's absence, Central Snoopers engineered a coup. They put a particularly vicious McDuff's Delivery Hound in the Presidential Palace, at the head of a government of well-trained Short Order Cooks. An arrest warrant was out for D'Arcy for obtaining money under false pretences and bad taste; he had put on forty productions of The D'Arcy Versey Story in one season. His bank accounts had been frozen.

We learned about our friend's difficulties from the twins' mother, Gladys Spandau. D'Arcy called her, asking for two million dollars to pay off his debts. She responded with a cake containing a hacksaw and a file. She did, however, tell Orville the sad news and asked him to call on her ex-ballroom dancing teacher when he was next in Buenos Aires, which was a week later. Orville was reluctant to help the failed showman, but he always tried to carry out his mother's wishes.

We put into Buenos Aires during Carnival and were swept up in the excitement of the festival.

The city had become Carnival Capital of South America after the government of Brazil forbade every kind of lewd entertainment and the Rio Carnival was no more. The Seven Day Vegetarian Church had taken the country over through their network of Born-again Chilli Peppers. They forbade beef production, turning the land over to corn and vegetables. Following the doctrine of 'If you enjoy something it must be sinful.' the Veggies banned all music, dancing, theatricals, meat, fish, alcohol and, of course, Carnival.

Gladys Spandau had told Orville that D'Arcy would be in disguise, dressed as an exotic female dancer. We watched the Carnival parades and eventually he spotted us.

"Thank goodness you have come to save me," said D'Arcy, swaying lasciviously. He always did have a willowy figure, but his female attributes were strictly limited: he looked like a skinny boy with an identity problem.

Our party stared at him in disbelief. I was the first to get to the nub of the problem.

"I gather from his brainwaves," I said, "that D'Arcy is without a fixed abode and is living on his wits. Dressed as he is, he is in constant danger of being arrested. I think we should assist him."

We did a conga, with D'Arcy in the lead, back to the Marine Explorer. A port guard approached him, and was immediately swept into a passionate, energetic tango, which eventually left him exhausted, sitting on a bollard. Back on the ship, Frank called Captain Jones to the saloon. He looked at D'Arcy.

"Where did ye spring from, my pretty?" said Jones/Ahab. " If I had my leg, I would ask ye for the next dance. Do ye come here often?"

"Captain," said Frank. "Mr D'Arcy is in disguise and needs to leave the country as soon as possible. Can we go?"

"We sail at midnight," said Jones. "Tomorrow we hunt the whale."

Exit Captain Ahab.

We sailed up the coast of Brazil taking a census of species and numbers. I was much better at this than Frank, because I could interview the larger species, which lived off the smaller fry. They had a very good idea if the fish populations were decreasing or not, because they ate them. They had noticed that factory ships had started coming into their waters and wondered why. Coco came up with the answer: McDuff's had bought the fishing rights to ensure supplies for their fish-burgers. The Brazilians were now

Fundamentalist Vegetarians (on pain of imprisonment or worse) so they did not require a fishing industry.

We entered the Bay of Rio. The setting sun bathed the mountain, topped by the giant statue of the First Cabbage, in a rosy glow. Rumour had it that a different statue had once been there. The statue of the First Cabbage contained a Vegetarian church and a revolving restaurant. That evening, we ate a delicious vegetarian meal, cooked by captured Buddhist monks, as we watched the 360-degree panorama pass before our eyes. We were able to appreciate the wonderful, skyscraper-studded bay, but the shacks made of wood and corrugated iron on the steep slopes reminded us that all was not well in the country. The waiter, who saw us staring at the squalid slums, warned us against going anywhere near them. According to him, terrible things happened there, including meat eating.

Among the many ships riding on the azure waters of the bay, Coco noticed a familiar shape. "That's the MV Couch," he said. "It's in Rio bringing compassionate care to the bewildered and befuddled. Heinrich Graft and Gerbil Sneed are on board. Do you think we should pay them a visit?"

Frank was against the idea.

"I don't trust that pair of crooks," he said. "Like all criminals, they are extremely moral. So I wouldn't put it past them to shop D'Arcy on the grounds that he puts on lewd theatricals. We are in Brazil, after all."

I was tuned into the Couch and caught a conversation between Heinrich and Gerbil.

"Are we organising a food distribution in the Rio

slums?" said Gerbil.

"Of course, we are," said Heinrich. "It's all very well ministering to the bewildered and befuddled, but those people are hungry. And, I doubt the First Cabbage's dictum, 'man cannot live by bread alone'. When man is starving, psychotherapy is a bit of a luxury."

"Do you know, Heinrich, that most of our budget is now spent on famine relief, clinics and low cost housing schemes?" said Gerbil.

"I do the cheques," said Heinrich. "And it was me who asked the Mother Superior to allow the nuns to run clinics. After all, they are medical doctors."

I couldn't believe my listening apparatus. Heinrich and Gerbil had changed from monsters of crime into champions of the poor. I told my friends and we decided to pay them a visit.

"They must be planning something really big,"

said Frank.

The Mother Superior greeted us cordially. She recognised D'Arcy, who had given up cross-dressing.

"I saw your production of the Sound of Music in Panama," she gushed. "It was wonderful. You must sign our visitors' book. It's not every day we have a star on board."

"You're too kind and I agree with your every word," said D'Arcy with his usual modesty. "I believe you are caring for a member of my company."

"Yes," said the MS. "Carmen Suarez is with us. She starred as Eva Peron in your production of *Evita*. The rejection was too much for her. She became extremely bewildered and occasionally befuddled, but she is making good progress."

"Thank you for helping her," said D'Arcy. "Let

me know when she is better. I'm thinking of putting her into my next production, The President Dubya Story. It's a crazy comedy about an illiterate idiot who made it to the top; it should cheer her up."

"It sounds like the male equivalent of *Evita*," said the MS. "Do you think the audience will believe it?"

"They'd better believe it," said D'Arcy. "He was responsible for the rise of the Vegetarian Church. He did more for vegetables than any president before or after him."

"We have another famous patient," said the MS. "General Quagmire is in the secure, padded accommodation. Slam Duncan, his assistant, brought him here after he tried to put Pot of Yoghurt in the White House."

"How are Sisters Agnes and Greta?" said Angela. 'Do they still run the Bank without

Frontiers?"

"They do indeed," said the MS. "But I must tell you, I have persuaded them that they are in fact male and should exchange their habit and wimple for a monk's robe. Come along. I'll take you to see them."

Orville and Frank frowned, D'Arcy looked scared and Angela smiled in anticipation. Coco and I were the first to move. We had to see the new Heinrich and Gerbil. We entered the BwF office to be greeted by a smiling pair of monks.

"Dr Frank!" exclaimed Gerbil, "It's good to see you. For a long time, I've been wanting to apologise for sacking you."

"That's OK, Dr Sneed," said Frank magnanimously.

"Brother Gerbil," said Sneed. "And this is Brother Heinrich."

"Greetings," said Brother Heinrich, "You are all

looking very well. Angela, I hope you have forgiven me for my past boorish behaviour."

"As a matter of fact, I haven't," said Angela, who was feeling nauseated by all the unctuous goodwill that was gushing from the Brothers.

"And D'Arcy," continued Heinrich, "I was so sorry to hear that you had been deposed from the Panamanian Presidency."

"No sorrier than me," said D'Arcy. "And now they have frozen my accounts and put an arrest warrant out for me. I'm so worried about my wife and son."

"Worry no more," said Gerbil. "We handle the McDuff's Slush Fund. I can arrange for a pardon and restitution of your cash. I have got so much dirt on McDuff's that they would not dare to risk being blackmailed."

Heinrich was thoughtful.

"I'm so pleased that we can help you, D'Arcy,"

he said, tears welling. "I know what it is to lose one's wife and baby son. He was only six months old. Thanks to the MS, I have come to terms with my loss and no longer feel the need to punish society by breaking the law."

For some reason, Frank was staring at Heinrich.

"What happened to your wife and baby son?" he asked.

Heinrich looked at Frank. There was something about his hooked nose and tombstone teeth that reminded him of his wife.

"I can now recount that traumatic part of my life without becoming homicidal," he said, and proceeded to tell the sad story of the fatal avalanche of beans and the runaway motorised shopping cart, which carried his baby into the San Francisco night, never to be seen again.

Frank was deathly pale. "Did you try to find him?" he asked.

"Try?" said Heinrich. "I moved mountains and fifty-seven agencies in my quest for baby Hans. In vain! I descended into a black pit, from which I have only recently emerged. I still have a can of beans with the lost person ad on it."

He opened his desk drawer and took out a can, which he handed to Frank. "There's a photo of Hans on it," he said.

Frank looked at the photo, which was not dissimilar to his baby photo taken by the orphanage. He dismissed the possibility of an amazing coincidence.

"All babies look the same," he said aloud. "Would you mind giving me a blood sample, Brother Heinrich? I'll contact my friends at the National DNA Registry. They might be able to help you."

Chapter Twenty

When we got back to San Diego, via the Panama Canal, we were reunited with Wilbur, Mary and the Delightful Dim Sums, who came to meet us. Frank had some leave coming, so we all took the MagLev to Culver for the Christmas holiday. The Ocean Explorer's voyage had only confirmed what we already knew: that the oceans were in a sorry state. We had only found a few traces of coral reefs. Fish populations, together with the larger species and mammals that depended on them, were in decline. The only ray of hope was that the deterioration was slowing down. Frank sent a specimen of Heinrich's blood to the National DNA Registry in Washington and then forgot about it.

We were in for a surprise when we got to the

Spandau estate. On the advice of Q, Coco's cucumber deputy, Wilbur had moved CoCoCo to Culver. In the process a good number of staff had been shed, which provided openings for local workers. The house had become the administration building and the labs were placed among trees in the grounds. The biggest surprise of all was the garage and annex, which was a reproduction of the one that was burned down. Orville immediately took up residence and rarely emerged.

The labs were gradually bringing out a range of drugs, which were going to the poorer countries at cost. Drugs to treat obesity in the rich world were financing those to treat malnutrition in the poor countries. We all settled into a new house that Wilbur had built for his family. The Delightful Dim Sums were learning not to move round the house in a series of back flips and forward somersaults. Wilbur hired a trainer for them to inculcate the finer points of gymnastics. In years to come, they would represent their adopted country at the Olympics, where they would to make a clean sweep of the medals. The Mongolian Government would protest at their defection and maintain that the glory should go to their native land. A compromise would be reached and the Stars and Stripes and My Mongolia would be played alternately at the medals ceremonies.

Frank received the result of the DNA test on Heinrich's blood sample. The National Registry had checked it against the records of orphans, and found that he was the father of Francis Frank, who had been abandoned in a shopping cart, at the age of about six months, near the

Golden Gate Bridge in San Francisco. Frank told us the news without any sign of emotion.

"Well, Frank," said Angela, "how do you feel about having Heinrich Graft as a father?"

"It's nice to know that I wasn't deliberately abandoned," said Frank, "but I can't say I like the idea of being his son."

"Will you tell him?" said Mary.

"Of course," said Frank. "I'll give him a call some time."

D'Arcy Versey returned to Panama and resumed his theatrical endevours. His bank accounts were defrosted and he was able to resume a normal life: but not for long. McDuff's, tired of the erratic behaviour of the incumbent president, one of their Delivery Hounds, gambled on the chances of his opponent, Arnold Swede winning. He lost and McDuff's paid the price. They lost the Panama Canal, which was awarded as first prize in a competition to find the longest palindrome. The winner was Runner Bean, who came up with: *a man a plan a canal panama.* Try it backwards and you will see how devilishly clever a runner bean can be. McDuff's were out of favour, so Central Snoopers ejected the Short Order Cooks from the Palace and installed a government of Spring Salads. The old charges against D'Arcy were reinstated, so he quickly put together a touring company of *No, No, Nanette*, which resulted in *Tea For Two* becoming the hit song of the year. Ever ready to take part in humanitarian efforts, the Shrinks without Frontiers ship was put at their disposal, and transported the company to San Diego, where they obtained political asylum. It was the first example in history of an asylum obtaining asylum.

D'Arcy paid a courtesy call on us a week before Christmas with a handful of tickets for his latest smash hit. Reunited with his wife and son and having obtained asylum with his bank balance intact, D'Arcy was a happy man. He became quite sentimental when I told him that Frank had found his father, but was loath to contact him. Ever one to interfere, D'Arcy couldn't wait to go to San Diego, go on board the SwF ship, and tell Heinrich he was a daddy.

The Brothers were in their counting house, e-mailing orders for humanitarian supplies for Panama, which had not recovered from the bombing administered by Central Snoopers. The new government, Spring Salads, hadn't the faintest notion of how to run a country. The Vegetarian Church had flooded the place with

Fundamentalist Vegetables, who were closing down butcher's shops, dairies and chicken farms and burning down all the branches of McDuff's. The shortfall in food supplies was causing widespread starvation. The new Heinrich greeted D'Arcy with a friendly hug.

"There is too little hugging going on," he said. "How can I help you, my friend?"

"Have I got news for you," said D'Arcy with relish, "Little Hans is found. He may no longer be a bouncing baby, but he is still your son. He is alive and well and residing on the Ocean Explorer, currently in San Diego."

Heinrich was transfixed. "His name?" he said, fearing the worst.

"No," said D'Arcy, "it isn't Captain Jones. His name is Francis Frank. The same Dr Frank you have known for years, and who you were trying to kill at one time."

"My God!" exclaimed Heinrich. "That's an argument against capital punishment, if ever there was one."

"Your son is in Culver at the moment. Will you come with me to meet him?" said D'Arcy.

Heinrich didn't answer for a while. "I won't go," he said. "He knows I'm here and he hasn't got in touch. He obviously doesn't wish to contact me so I must respect his wishes."

D'Arcy left the MV Couch, thinking how stupid some people could be. Heinrich had a lot to be ashamed of, but that was all in his past. As the founder and financer of Shrinks without Frontiers, he was eligible for the Nobel Peace Prize. Not every waif and stray could boast of such a father.

We all gathered at Culver for a Christmas celebration. Frank and Angela were there. Since they had not been given the Doomsday gene like Coco and me, Angela was expecting a son for them and a grandson for Heinrich. Brothers Heinrich and Gerbil arrived, because Frank had decided that, in view of his impending paternity, it was only fair to tell his father. The pair came bearing gifts: a gold bracelet for Angela, a bottle of aftershave for Frank, and a voucher for seventeen years education for the unborn child, Hans. Orville had invited Captain Jones, who arrived with D'Arcy and his family. The Captain's schizophrenia was worse: his personality now wandered between Captain Ahab, Napoleon Bonaparte and King Louis XVI of France. When he wasn't ranting on about the whale, he was cursing Wellington, or hiding in the cellar from the Parisian mob. Juan, the family retainer, was always embarrassed when Frank visited. He could not help remembering the passionate declarations of eternal love that he had made to Teresa Martinez.

We had a Christmas party for which D'Arcy organised the entertainment. He and his wife, Conchita gave a brilliant exhibition of Latin dances. Captain Jones gave a reading, which jumped from Melville's Moby Dick, to Napoleon's speech to the Old Guard before Waterloo, to Dickens' Tale of Two Cities. Heinrich read the Dickens Christmas favourite, and The Delicious Dim Sums put on a wonderful tumbling act which climaxed in their building a human Christmas tree, complete with lights and shiny baubles and a star at the top.

Coco and I observed the festivities with amusement. Having no requirement for food and drink, we sat at table and told funny stories to amuse the Dim Sums. Over the holiday, Coco and I had made a decision: since we were immortal, it was not a good idea to become too attached to mortal creatures. We saw how distressed the Dim Sums were when their pet rabbit passed away. We decided to retire to a remote place, where we could meditate on the transience of everything except us. Coco had been studying Zen Buddhism for some time and suggested we find the remote Himalayan Valley, which he had read about in an ancient novel. All our friends were sad but understanding when we told them of our decision.

So, as I mentioned at the beginning of my story, I am sitting by a rivulet, dangling my feet and prehensile tail in the cool water, narcissistically admiring my reflection. Coco and I trekked through mountain snow and ice before we found our valley. It isn't the one mentioned in the novel; we found that had a Hilton Hotel and Ski Resort, so we moved on. Coco and I have shut down a lot of our equipment. We have grown rather tired of history repeating itself. From our study of the history of the human race, we came to the conclusion that man had made very little intellectual progress since the Stone Age. He is forever trying to bomb his fellows back to that era. Most worship the Great Cabbage, which we find extraordinary. Some have gone back to the ancient religion, which is just as extraordinary. It is true that science has taken great strides in expanding human knowledge, but the majority of the monkeys-without-tails display a singular ignorance of what is really important in life. Coco and I have found it: peace.

Part Two

Chapter Twenty-One

The final word of Part One was *peace*. Coco and I really thought we had found it in our valley. Time passed, I can't say how much, because, being immortal, Coco and I had stopped counting. We had closed down some circuits, which were of little use to us and were concentrating on important problems such as the meaning of life, and how to preserve the flora and fauna of the Earth. The solution we came up with for saving the planet was to make the monkeys-without-tails extinct. After all, they were already doing their best to bring this about. But we came up against a moral problem: we were unable to contemplate such a radical solution because we were very fond of some of them. We still remembered the Delightful Dim

Sums, Frank, Angela and little Hans. Orville and Wilbur Spandau and Mary were often in our thoughts and we never failed to send them birthday greetings. We did our best to make a contribution to saving the world, but I'm afraid we did not achieve much.

While we were in our valley, we designed an anti-gravity motor, which was first manufactured by CoCoCo in Culver Town. This made a great contribution to cutting down pollution, and, after it came into general use, sea levels only rose five centimetres a year. The improvement was not better because rich people could not bear to give up their SUVs. Arnold Blackpudding, the Governor of California, set the trend by having a custom-built model with six wheels, which did one kilometre to a gallon. Coco sabotaged his brain, (it was not difficult) by emitting voices, which only the Governor could hear. He started speaking in tongues and became so popular that the constitution was changed to allow a foreign-born idiot to become president.

He walked into the White House on a landslide vote. The first thing he did was to fit powerful Anti-gravs to the main building so that he could go on foreign tours without leaving his office. Did that set a trend! Everybody who could afford them fitted Anti-gravs to their residence and went on vacation in it. Some employees started going to work in the house and "I couldn't get the house started" became a common excuse for tardiness. To satisfy the SUV maniacs, we designed a model, which was operated and sounded like the real thing. The Anti-grav motor allowed it to travel on the road, off the road, on the water and in the air.

When we first arrived in our valley, Coco and I were completely alone, except for the flora and fauna. We used to have long conversations with the goats, which scampered up and down the mountains, pausing occasionally to clash horns with a rival. We persuaded them that marriage counseling was a more civilized way of settling family disputes. Eventually, they took our advice and talked over their problems with their rivals. When they realized that they were getting fewer headaches, they were pleased that they had listened to us.

The rot set in when a hermit came into the valley and took up residence in a cave. We thought there was something not quite right about him when he had all the comforts of civilization flown in by Supplies-R-Us helicopters. They airlifted a gang of craftsmen to the valley, who transformed the cave into a desirable one-bedroom dwelling with en-suite bathroom, living room and dining kitchen. Solar and wind power ran an array of communications and home entertainment equipment. You could set your calendar by the weekly helicopter drop of essential supplies like Patagonian Tooth-Fish and Champagne.

Coco and I were not at all pleased by this development. We climbed to a ridge overlooking the cave and inspected the hermit. He was wearing a brown smock with a hood, which had Balenciaga written all over it. His leather sandals were obviously Gucci and the cord round his waist with diamond-studded cabbage buckle was by Cartier. In spite of these luxurious accoutrements, there was no mistaking him: it was none other than Brother Gerbil Sneed. Coco and I were shocked; the last time we had seen him, he had been enjoying the Christmas party at the Spandau residence. We thought he had settled down to running the Bank without Frontiers on board the MV Couch. We resisted the temptation to approach him, preferring to spend the next few hours establishing the reason for his move to our valley. The facts turned out to be rather grizzly.

One Monday morning, Brother Heinrich returned to the ship after spending the weekend with his grandchildren. Hans was a lively three-year old and he now had a baby sister, Gretel. Heinrich doted on them but his relationship with his son was not easy. He had, after all, almost had him killed. As in a Greek tragedy, this thought was never far from his mind when he was with him. Brother Heinrich went straight to the counting house and was shocked to find that Brother Gerbil had his right hand swathed in bandages.

"What happened?" he asked, troubled by Gerbil's pallor.

"Luigi Parmesano was here," he said. "He arrived with a particularly vicious Delivery Hound."

"I thought McDuff's phased them out, when they got a taste for fingers."

"So they did," said Gerbil. "But people like Luigi snapped them up to use as enforcers."

"What did he want?"

"I did a really stupid thing," said Gerbil. "I used a few million from his account to finance flood relief in Switzerland."

"Well put the money back in his account," said Heinrich.

"I can't," said Gerbil. "There isn't any spare cash in the bank. Every cent is committed to some wonderful project or other."

"So what did Luigi say?"

"Put your hand in the Delivery Hound's pouch and take the letter."

"Did you take the letter?"

"I got hold of it, but I let it go"

"Why did you let it go?"

"Because it bit my finger so."

"Which finger did it bite?"

"The little finger on the right."

Gerbil held up his mangled hand. "I only have four fingers on this hand now."

"We'll re-attach the finger," said Heinrich. "It's child's play for our surgeons."

"The Hound ate my finger," wept Gerbil. "I won't be able to play the guitar now."

"But you couldn't play it before," said Heinrich.

"I was going to learn."

"Use the Django Reinhardt Digitally Challenged Method," said Heinrich. "But the question remains: how are we going to placate Luigi?"

"If I don't put the money back in his account," said Gerbil, "he says he's going to come back for an arm and a leg."

"An arm and a leg!" snorted Heinrich. "The Don has no imagination. Even his crimes are clichés. There's only one thing for it: we'll have to disappear for a while. I'm going to close the bank for stocktaking and spend more time with my family. That will allow fresh funds to come in. What are you going to do?"

"I've found a remote valley in the Himalayas where there is a suitable cave available. I have been thinking about going into retreat for a long time. I would like to contemplate the eternal truths of the balance sheet of life and find peace."

"I wish you luck," said Heinrich, "but it sounds to me like you are dropping out. What about the bank? Who will do your job? I certainly can't; I have more important things to do, like dandling my grandchildren on my knee."

"I've been in touch with Yodel Lay in Panama," said Gerbil. "He's ready to leave, because Durian Fruit, who holds the presidency, stinks, and nobody is willing to go near him, so nothing gets done. He thinks Central Snoopers are about to make a move on the country to replace Durian Fruit with a woman, Rose Bouquet. She can't do any worse than Durian Fruit, and she'll smell a lot sweeter."

"If that's your decision," said Brother Heinrich, "I'll have to accept it. But I will be sorry to lose you." He hugged Brother Gerbil. "There's not enough hugging in the world," he said, with a tear in his eye.

"Well, Coco," I said, "now we know why Gerbil Sneed is here, what are we going to do?"

"Ignore him. He may have changed, but I have never liked him."

It proved impossible to ignore Brother Gerbil. Apparently there was an exodus of African dictators to fresh fields and pastures new, laden down with loot from the Treasuries of their failed states. This found its way into the Bank without Frontiers. Luigi Parmesano was compensated and lost interest in Brother Gerbil's arm and a leg. Relieved at not having to contemplate the balance sheet of life from a wheelchair, Brother Gerbil began broadcasting his philosophical insights to an expectant world in the *Brother Gerbil Golden Hour*. Before long, geriatric devotees began to turn up in the valley on the weekly Supplies-R-Us flights. They insisted in giving all their money to the Golden Hour Trust and taking possession of all the available caves, which had been made into comfortable homes by a new company, Caves-R-Us. Chanting started at five in the morning and went on most of the day, interrupted only when some pensioner slipped and broke a hip. All this was too much for Coco and me: we decided to leave.

Chapter Twenty-Two

General Quagmire was in therapy with the Mother Superior.

"Slam Duncan should never have brought me here," he was saying. "There is absolutely nothing wrong with me."

"What about the pot of yogurt that you wanted to install in the Presidency?"

"He's no worse than the present incumbent."

"I agree, but the president is a living entity."

"So is yogurt."

"And what about your obsession with watching your shadow, even to the extent of walking backwards when the sun is in front of you?"

"I'm over that."

"And you have fitted rear-view mirrors to your eyeglasses."

"It solved the problem. Now sign me out, please."

"Very well, General. I'll call your assistant and ask him to collect you."

The MV Couch was moored at Los Angeles, distributing food and organizing shelter for the homeless. In spite of being the richest city in the world's richest country, there were hundreds of street sleepers. The Mother Superior stopped comforting the bewildered and befuddled in the poorer section of the community, and concentrated on the rich and powerful. They had cut themselves off from the harsh reality of poverty and exposure to the elements in their walled estates, protected by armed guards and ex-McDuff's Delivery Hounds. The Mother Superior risked life and fingers ringing the bells of the rich and famous. She managed to get a few minutes with the movers and shakers of Los Angeles. She found the film stars most susceptible to hints of hellfire if they did not cough up a large contribution to the Sisters of Commiseration. Most of them felt guilty about the things they had done to get roles in the first place, and the millions they got for making lousy films in the second place.

Slam Duncan turned up at the MV Couch to

collect his boss. Slam, at nine feet tall, was one of a batch of genetically modified sportsmen who had been given a growth gene. The experiment was so successful that nobody less than seven feet tall was allowed to play in the NBA. Slam had a fairly successful career, which was brought to a stop when he drove under a low bridge in his coupe and cracked his skull.

"Finally!" said the General. "Am I glad to see you, Duncan. Get me out of here. I'm sick to death of hearing myself talk about my childhood, my relationship with my mother and listening to selected readings from the works of Sigmund Freud. The Mother Superior has established that I am suffering from every condition known to psychiatry which, I believe, fits me for an elevated position in the military hierarchy."

"It's good to see you, General," said Slam. "I've

got your office outside. You'll be able to slip straight back into your old routine."

There was a parking ticket on the General's office when he reached it. Slam started the Anti-gravs and the building rose smoothly to ten thousand feet and cruised to the Central Snoopers complex near Washington. On the way, they passed the White House, which was heading for Camp David, which was on its way to Moscow to collect the Russian president, Boris Borscht. The current American president was Sam W.Yam, a six-foot tuber of African-American descent. Passing the White House reminded the General of the sad demise of Arnold Blackpudding, the previous president.

As the Governor of California, Arnold had been able to spend a lot of time with his ex-colleagues in the film industry. He visited the studios regularly, slapping a male back here and other female body parts there. He had regular sessions with his dialogue coaches, budding starlets, who were about to be cast in mega-movies. They had been about to be cast in a star roles for a long time, but had been landed with the task of helping Arnold with his elocution. No matter what they did, he always emerged from the classroom caravan, perspiring slightly and speaking in exactly the same Scandinavian tones, so beloved by his fans. When he became president, his gallop was strictly controlled. Security forbade him the delights of Tinseltown and his SUV was taken off the road.

"I'm the President!" shouted Arnold at the Secretary of State, Kohl Rabi. "If I want to bomb Cambodia, I'm allowed to bomb it. If I want to invade that other country, I can invade it. But if I want to take my SUV out for a spin, I can't spin it."

In the end, Arnold was allowed to fly his SUV, now fitted with Anti-gravs, to the Arizona Desert. To justify the excursions, he was always accompanied by one of his dialogue coaches and a picnic lunch. He relished going over rocks and dunes and along perilous mountain trails. He never had an accident because his SUV was fitted with collision avoidance systems. He could run at a cliff face at a hundred miles an hour, but the SUV would come smoothly to a stop before a crash could happen. This really annoyed Arnold. If he could not crash, there was no danger. If there was no danger, he was bored. He disabled the collision avoidance systems and proceeded to take out a series of huge cacti. Being somewhat intellectually challenged, if not downright stupid, Arnold forgot he had disabled the collision avoidance systems and raced at the edge of a deep canyon, the Grand Canyon, no less, afterwards renamed the President Blackpudding Canyon, and sailed over the edge. The shock had a drastic effect on Arnold, who said his last words in perfect English: *The rain in Spain stays mainly on the plain.* His companion ejected and landed safely by parachute. Arnold was so amazed by his sudden command of English pronunciation that he was too slow to operate his ejector seat. His companion's reaction to Arnold's sad demise was, "By God, he's bought it!"

General Orpheus Quagmire contemplated the ephemeral nature of man, as his office arrived at Central Snoopers. He stepped out to be met by an honour guard of six-foot yams. President Sam

Yam was a great believer in nepotism, which he said had shaped the history of America. The honour guard was composed entirely of the president's relatives and threatened to bankrupt the treasury with its demands for fertilizer and other perks. General Quagmire was pleased to be welcomed in this fashion. On one occasion, the honour guard had been composed of Dwarf Beans. This upset the general, who believed that size really did matter. Slam Duncan led the General back into his office.

"Now, Duncan," he said, "the time has come to get those folks who sent me to the funny farm."

"Yes, General," said Duncan. "I have compiled a list pending your return. Spud Murphy, the Union man, was the ringleader. He got the whole potato crop to go on strike and eventually migrate to Ireland. This led to the Great Boston Potato Famine. Hundreds of policemen lost weight and were too weak to patrol the streets. A huge crime wave gripped the inner cities. The president, himself a distant relative of the Murphy clan, had to sign the committal papers. As a cover, he put out the story that you wanted to install Pot of Yogurt as president."

"But I did want Pot of Yogurt in the White House," said Quagmire.

"You didn't really, did you, General?"

"Didn't I? Oh."

"As you can see, General," said Duncan, "your in-tray is empty."

"What does that mean, Duncan?"

"It means that you are useless, unnecessary, superfluous and redundant, General."

"Duncan! This is gross insubordination. I'll have you court-martialed."

"Too late, Quagmire. You're a hopeless case. I'll demonstrate why you are ready for the scrap heap. How tall are you?"

"Six foot, two."

"General, you are five foot one and a half."

"Rubbish. Here's my yardstick. Measure me."

"You sawed six inches off your yardstick, General."

"I think I am six foot, two, therefore I am six foot, two."

"You're putting the Descartes before the horse, General. Carry on like that and it will be back to the funny farm instead of that nice retirement home in Guam."

"I'm not retiring and that's final."

"General. You are retiring and here is your successor."

Slam Duncan slapped a rotten apple on the General's desk.

"Meet General Cox Pippin."

"He has a smell of corruption."

"Cox Pippin is rotten to the core. Just the apple we need at the head of Central Snoopers. President Yam has appointed me Cox Pippin's deputy. To all intents and purposes, I shall be in charge of the day-to-day operations of the company. Air Force One, complete with padded cell, awaits you."

With that, the honour guard, pistols cocked, burst in and marched the General out of the office.

Chapter Twenty-Three

President Sam W. Yam was preoccupied. He was on his way to Africa in the White House and he hadn't the foggiest notion which countries were in Africa, who their presidents were, even where Africa was in relation to Washington DC. The doorman put on the *Fasten Seatbelts* sign, and took the White House up to twenty thousand feet to avoid turbulence over the Great Lakes. Sam was a handsome, robust yam. His eyes were piercingly clear. His skin was smooth with a tinge of red, no doubt due to exposure to the sun. His topknot was neatly trimmed crew-style. His arms were strong from pumping iron and the two tendrils, which he used for walking, were excessively muscular. His only regret was that he had no chin. Sam liked to keep fit and went jogging every morning. He took a sip from his morning glass of nutrients, pulled a face because it tasted like diluted dung, which it was, and pulled on his tracksuit. A sharp jog round the White House grounds was just what he needed to start the day. He opened the front door, walked out and fell twenty thousand feet.

The Bushman family was sitting round a fire brewing tea when President Yam dropped in on them. Dan Bushman raised his cutlass, about to lop a piece off the six-foot yam.

"Steady with that thing," said Sam. "You might hurt somebody."

"Another of the talking yams," said Dan Bushman to his wife, Dana. "The way things are being genetically modified, there won't be any food left that doesn't talk back and argue about the best recipe for cooking it."

"I don't fancy it," said Dana. "Those huge yams are woody and tasteless."

"I'd better introduce myself," said Sam. "I'm Sam W. Yam, President of the United States. You must have heard of me."

"I'll say we've heard of you," said Bart Bushman, the eighteen-year-old Bushman son. "Your sayings are famous. If we want to get people rolling round, helpless with laughter, we read them aloud. I've worked out a stand-up routine using the material. The tourists love it."

"You get tourists here?" said Sam looking round at the collection of miserable mud huts. "Do they stay here?"

"Keep it up," said Bart. "I'll get my notebook."

"No," said Dan, indulgently. "They stay at the Kalahari Hilton up the road or at the Namibia Sheraton an hour away. We live in condos in the model village and only come to this dump to lend the place a little local colour. It's painful. You should hear the stupid tourists, 'Just look at those little people. Aren't they cute? Do you think we could take one home with us?' and so on."

Sam noticed that Dan had a bow and arrows.

"Do you hunt and gather?" he asked.

"The bow and arrows are more local colour," said Dan. "I do a bit of farming and keep a few goats and sheep. We don't do any hunting and only gather our monthly dividends from the mineral rights."

"I gather the desert melons," said Dana. "They are small and aren't really melons, but they quench your thirst."

"Tell me," said Sam. "Why weren't you surprised when I fell out of the sky?"

"It happens all the time," said Dana. "We have those MonCul yams which can move around to find water and good soil. They are objecting to being dinner and have started climbing trees. They wait for us to walk under them and drop on our heads. Some of them weigh a ton, and can do a lot of damage. Now you tell me how you managed to fall twenty thousand feet and not finish up as pureed yam?"

"I have sensors," said Sam, "which release a parachute when I fall from a great height, and air bags, which open just before I hit the ground."

"Of course," said Dana. "Sorry I asked."

"That's all right," said Sam. "Do you think I could have a cup of that tea?"

"You can't drink that filthy stuff," said Dana. "It's just more local colour. We're going for a coffee at Megabucks."

"Of course," said Sam. "Sorry I asked."

"Right," said Dan, "come on then. We're on our break. We'll take you to Megabucks. Our treat."

As he drank a latte, sitting comfortably in the Kalahari Megabucks, Sam enquired about the local telecommunications setup.

"You can e-mail, fax, phone, text, video conference or use the talking drums," said Dana. "I don't recommend the drums, because they charge a fortune and only have a range of five miles. Use my mobile."

Sam got in touch with the Secretary of State in

the White House, which was just about to land on Table Mountain, Cape Town.

"Where the hell are you?" screamed Kohl Rabi, the Secretary of State. "We were just about to announce your death and the appointment of Slam Duncan in your place."

"What do you expect?" said Sam, "It's easy to forget that the White House can fly. Can't you lock the doors or something? And I'm disgusted that you should think that big oaf, Slam Duncan, could replace me."

"Watch your step, Sam," said Kohl Rabi, with a hint of menace in his voice. "A nine-foot basketball star will always have it over a six-foot yam. So don't get uppity."

"Where are you, anyway?" said Sam.

"We've just landed on Table Mountain, Cape Town, South Africa," said Rabi. "Take out the crib I made for you and learn the relevant facts by heart. I don't want you starting off with 'It's great to be with you Africanians, here in the heart of Africania'."

"I could never make a mistake like that."

"You already have. I'm quoting you. Where exactly are you? I'm getting a Kalahari copter to come and pick you up."

"I haven't the faintest idea."

"Use your GPS."

"I haven't got GPS."

"There must be folks there. They'll have GPS, you moron."

Dana came to Sam's aid with the relevant coordinates, which he passed on to the Secretary of State.

Kohl Rabi slammed his phone down. He turned to the Chief Presidential Adviser, Anthony Capote.

"I've had it with Sam W. Yam. He's got to go. We need somebody with just enough intelligence to do as he is told, and not go jogging at twenty thousand feet. Who do we have lined up?"

"We got Slam Duncan, but he too young; Luigi Parmesano, but he too old," said Anthony, who was still struggling with English, in spite of the fact that 'he born here'. "General Quicksand not got no job and live retired in Guam now he leave Central Snoopers."

"Are you by any chance referring to General Orpheus Quagmire, Anthony?" said Kohl Rabi.

"Same, same," said Anthony who had made a special study of synonyms.

"Get him on the phone. I want him in the White House by tomorrow."

It only took a few minutes to get General

Quagmire on the phone and tell him that a new and glorious vista had just opened up for him. Ten minutes later a feast, to be paid for by the US government, was announced to the Kalahari Bushmen. The special occasion would require all the yams of six feet and over. Pigs were killed, yams were slaughtered and both were made into puddings, wrapped in leaves, and baked in earth ovens constructed by the Kalahari Boy Scouts. Despite her reservations about huge yams, Dana Bushman declared the yam and pork puddings the best she had tasted since a Vegetarian Church Missionary received the same treatment. Alas, poor Sam W. Yam was no more, pleasing more people in death, than he had ever done in life.

Chapter Twenty-four

Coco and I went to visit Brother Gerbil Sneed

before we left our formerly peaceful valley, which now resounded constantly with chants, the tinkling of bells, the gonging of gongs and the clack of helicopters bringing in pop stars for a retreat in the vast Ashram, which had been built by the Bechtel Corporation of Iraq. Gerbil was sitting in the midst of a group of his followers perorating on the dangers of modernity.

"Having sloughed off the cares and obsessions of the material world, you will now be able to develop your spiritual potential and get in touch with your real self. If you are like me, you will not like your real self very much. Most of you have been selfish and grasping, building up treasure on earth, which you have now deposited with the Golden Hour Trust. Relieved of the burden of wealth, you can now enjoy the simple things in life: a walk in the mountains, a cup of milk provided by a mountain goat, and a loaf of unleavened bread you have baked in a biscuit tin oven." Gerbil noticed us and paused. "I would like to welcome two of the products of my previous life: Coco and Radish. They are a good illustration of the futility of modern science. Although they have the hardware and software to master the total knowledge of the world, they lack one important thing: a soul."

"Come off it, Gerbil," said Coco. "Don't be so patronizing. If you have detached yourself from the debilitating pleasures of the world, what about the weekly deliveries of Patagonian Tooth-Fish and the wine cellar you have in the back of your cave?"

"I have those things to treat a medical condition," said Gerbil. "Now if you don't mind, I will continue my peroration."

Coco and I left without a word. Whatever Gerbil

was up to, he was enjoying it. As we made our way to the Golden Hour Airport, built to facilitate the arrival and departure of Golden Hour devotees, I experienced a feeling of relief and anticipation: relief at getting away from the hypocrisy of Brother Gerbil and anticipation at the thought of seeing my friends in Culver and San Diego.

Culver had changed radically in the years we had been away. The development of the Anti-grav motor had brought a flood of work and wealth into the town. Some of the new houses made the Spandau residence look like an outhouse. There were swimming pools and Anti-gravs everywhere. Even the kids had Anti-grav scooters and bikes. The population of Asians had soared as the rich residents employed domestic helpers, gardeners and chauffeurs, and the

Philippines had been largely depopulated. The Delightful Dim Sums were now about ten years old (they didn't know their precise age) and had just made a clean sweep as the American gymnastics team in the last Olympics.

We visited Frank and Angela in San Diego, where Frank was working for MarineSurveyCorp. Little Hans was in Kindergarten and having a problem with accepting Gretel, his new sibling. He had already tried to fit an Anti-grav to her stroller in the hope of sending her back to where she came from. Grandpa Heinrich visited every weekend, leaving the management of the Bank without Frontiers in the capable hands of Yodel Lay. The latter had finally officially tied the knot with Friday Knight and the happy couple were living in perfect domestic harmony on the MV Couch.

Coco and I basked in the warmth of our friends' families. When we visited Frank and Angela, in San Diego, we noticed that baby Gretel's room was equipped with a baby alarm, which simply broadcast sounds to a receiver. I considered this particularly unsatisfactory. Coco and I set about designing a transmitter, which would provide parents with specific information. We finished up with a light plastic bracelet, worn by baby, capable of sensing and automatically transmitting numerical descriptions to a receiver bracelet worn by a parent:

1 for Number one;

2 for Number two;

3 for Where's my bottle?

4 for You are starving me;

5 for My Teddy has fallen out of the cot;

6 for I'm about to climb over the side of the cot;

7 for I'm too warm;

8 for I'm too cold;

9 for There's somebody under my cot;

10 for My sibling is trying to suffocate me with a pillow.

The device swept the world and made the Spandaus even richer.

It was several weeks before Orville emerged from the garage to meet us. When he did, he had the complete designs for an underwater vehicle with several weeks' autonomy and room for two scientists, a captain and a cook. He greeted us warmly and asked us if we would like to meet his latest creation, a rabbit, which could jump over buildings as well as doing most of the things Coco and I did. We couldn't quite see the point of Rabbit's high jumping prowess, but it was to come in useful at a later date. Orville went off to San Diego to get his submersible built, leaving Rabbit in our care.

"Just wait until you see the sub," said Rabbit. "It can go deeper than anything before it. We'll be able to explore the deepest parts of the oceans."

"We?" I said "Is Orville taking you on his voyage?"

"He certainly is," said Rabbit. "He also mentioned that he was going to ask you two to go along."

"That's nice of him," said Coco. "But what can we do on a sub?"

"You do realize," said Rabbit patiently, "that between us we have an enormous, comprehensive range of programmes and abilities, which will be very useful at sea."

"Well," said Coco, "I can see one disability that you have. You need food. Radish and I are totally autonomous."

"What about your need for sunshine, Radish?" said Rabbit.

"An ultraviolet lamp does the job for me." I said.

"And I can survive on any kind of electronic radiation," said Coco.

I could see that we were going to have problems getting along with Rabbit, who was female, quite self-important, and a bit of a know-all. While the sub was being built, the three of us entered into the fun of our large adopted family. Rabbit, always a show-off, combined with the Dim Sums in a routine which had her leaping to the top of the human pyramid they formed at the end of their tumbling routine. She was equally adept at dismounts involving a back somersault with half twist, landing lightly and firmly on her back legs. If I had had a nose, I would have felt that it had been pushed out of joint.

Orville returned with the news that The Nemo, for that was its name, was ready for sea trials. I thought the name, which means 'Nobody' in Latin, quite inappropriate. Everybody and even every ship has to be somebody. But I let it pass; the learning of Latin had disappeared, a sad loss to human knowledge and a great relief to schoolchildren. Frank, the captain and the cook were already on board the Nemo. Orville, Rabbit, Coco and I were to make up the rest of the crew. We set off for San Diego, leaving Wilbur and his boisterous family behind. The possibility of the Dim Sums doing back flips in a submersible was fraught with danger, so they had to stay behind.

The Nemo was not, in fact, in the water, but in dry dock, so we were able to appreciate her wonderful lines. Built with the aerodynamics of a Giant Squid, without the tentacles, she was sixty feet long with a large girth designed to withstand the enormous pressure of deep dives. We went inside and discovered the commodious accommodation afforded by the broad dimensions of the sub. Coco and I joined Frank, who had settled into his cabin, which had a luxurious en-suite bathroom, and the rest of the accommodation was equally comfortable. The working area was equipped with the latest guidance systems, video equipment and everything the scientists needed to analyse and preserve specimens. We went into the saloon, where Coco and I renewed our acquaintance with the captain.

After the Southern Ocean expedition, Frank persuaded Captain Jones to commit himself voluntarily to the MV Couch, where the Mother

Superior took personal charge of his therapy. It didn't take her long to establish that he wasn't Captain David Jones, Captain Ahab, Napoleon Bonaparte or Louis XVI of France. He was, in fact, Hiroshi Nagasaki of the Japanese Imperial Navy. He had been in the Battle of the Diaoyu Islands, from which he emerged totally bewildered and seriously befuddled. As a fighter pilot, stationed on a carrier, he had been drafted into the 42nd Kamikazi Squadron (numbers 1-41 no longer existed) and ordered to crash his plane, filled with high explosives, onto the deck of a Chinese carrier. Ever obedient to the orders of Emperor Sukiyaki, he climbed aboard his flying bomb and headed for the nearest carrier, the Choufleur, a French ship on loan to the Chinese navy. The whole point of being a suicide bomber is not to survive, which would make a mockery of the whole business. Suicide bombers who boast of several successful missions are not to be believed. Hiroshi dived, but at the last moment, his training took over, and he made a heavy landing, covering the Choufleur in cheese

The fact that the carrier was covered in cheese needs some explanation, so here goes. The Parmesano Munitions Factory was just one of Don Luigi's enterprises embedded, for security reasons, in his processed cheese factory. After a heavy night celebrating the end of the spaghetti harvest, an employee confused the cheese and high explosive supply pipes, causing the packets for cheese sticks and party cubes to be filled with dynamite, and the munitions to be filled with soft cheese. The company sold the day's production of munitions, filled with processed cheese, to the Japanese Imperial Navy at a big discount, never expecting them to be used. Well, they were. Hiroshi crashed his plane, and the detonator pack exploded releasing a flood of rancid cheese. As the only suicide pilot to make a carrier into *choufleur au gratin* and survive, he felt he had become a figure of fun, derided and mocked, in his head, by the silent majority. (That means the dead, by the way.) The strain of being Hiroshi Nagasaki became too great for him. He adopted a new persona, that of Captain David Jones. The significance of the adopted name was not lost on the Mother Superior, who was aware of the connotations of 'Davy Jones' locker'.

Captain Jones emerged from the MV Couch, a new man: Hiroshi Nagasaki. He was in every way normal except one: he could never bring himself to order any gratinated dish in a restaurant; but then, the Japanese aren't fond of cheese anyway. At first, it was difficult to get used to calling him Captain Nagasaki or Hiroshi, and it was also difficult to get used to his bowing. He was constantly banging his head in the confines of the sub, so Orville ordered him to stop. When we put out from San Diego, Hiroshi demonstrated an unbelievable devotion to his duties and we proceeded without a hitch towards the Pacific trenches.

The final member of our crew was a French cook, Richard Trichet, who astounded us with the dishes he was able to turn out using only microwaves, since naked flames, steam and smoke were forbidden in the submersible. Richard, we called him Rick, was a charming middle-aged man. He had dark curly hair, a strong chin, a perfect nose, high cheekbones, a body-builder's figure, and, for a dark man, strikingly blue eyes. It was said that women swooned over him wherever he went. There was something about him that was familiar, but I couldn't put any of my twenty fingers in it. Perhaps he reminded me of some film star action hero.

Chapter Twenty-Five

The 'Fasten Seat-belts' sign went on in the White House. As it rose to 15,000 feet and headed for Washington, President Orpheus Quagmire gazed down at the Kalahari Desert and thought about his predecessor, who had recently been consumed as pork and yam puddings. He toyed with the idea of opening a chain of pork and yam pudding outlets, but dismissed the idea; he was going to be far too busy running the world's most powerful nation.

He got on the videophone to the former Vatican where the Cardinals were due to meet in a month's time. He was put through on videophone to Gloria Inexcelsis Deo, the diminutive Filipina, who was currently Chairman of the World Organization. She had been chosen out of sympathy for Pineapple del Monte, the former Chairman, who had suddenly rotted.

"Congratulations on your elevation to the presidency," she said. "I was so sad to hear of the fate of Sam W. Yam. We became very good friends when I was the President of the Philippines. He comforted me when my country was depopulated by the brain-drain to your country and my shoes were destroyed by mould."

"Yam died serving his country," said Orpheus, "and personally served the hungry in the

Kalahari Desert. I'm phoning to ask you about the next meeting of the Cardinals."

"It's on the tenth of next month," said Gloria. "The main business will be Mars and the unfinished business we have with the Red Planet."

"I can't stomach Communists either," said Orpheus. "But I really wanted to raise the matter of Slam Duncan, the effective Head of Central Snoopers. I believe he is a menace to society and doesn't know his place."

"What is his place?"

"He's a nine-foot basketball player and his place is on a corner lot teaching delinquents the finer points of the game."

"And he's black."

"That has nothing to do with my objections to him as head of Central Snoopers," said Orpheus. "He also has a bad habit of picking his nose with his thumb."

"That's impossible."

"I'm glad you agree," said Orpheus. 'Shall we vote him out?"

"You'll have my vote on one condition, Mr President."

"Which is?"

"You stop poaching our doctors, nurses, physiotherapists, opticians and all the other highly trained professionals," said Gloria. "The last time I went to the doctor's in Manila, there weren't any. I had to make do with a plumber."

"I'm sure he did a good job," said Orpheus, "Tell him we need plumbers as well. What about domestic helpers?"

"You already have all the ones who want to work over there," said Gloria. "I can tell you that it's causing major strife in my country. Grandmas are bringing up kids, who are running wild. Fathers are on the town every night. The social problems are enormous."

"We can send you some of our Blacks and Hispanics in exchange."

"I'll forget you said that, Orpheus. I'll see you at the next WO meeting."

Gloria checked the video recording of the conversation, and filed it under B for blackmail. "It should be worth a small aircraft carrier before the next election," she said to her assistant.

As the White House touched down in Rome, Orpheus put a tick against Slam Duncan's name. The second name on the list was Brother Heinrich Graft, the Chairman of Shrinks without Frontiers and Managing Director of the Bank without Frontiers. He called for the file on the MV Couch and her personnel. The ship was using San Diego as its main port, because

Brother Heinrich visited his son's family every weekend. The ship went out when there was a particularly nasty war and loaded up with traumatized victims. The main effort of Shrinks was now focused on famine and disaster relief. Being stationed in California made fund-raising easier, which supplemented the huge sums contributed as commission by the BwF. The bank's clients were especially prone to violent death and nobody ever came forward to claim the dear departed dictators' loot, which did not officially exist. These windfalls went straight to good works. Orpheus was particularly upset with Shrinks because they had allowed Slam Duncan to commit him when he was perfectly sane, and with brother Heinrich, who had done nothing to help him.

He turned to Secretary of State, Kohl Rabi,

who was penning a threatening letter to the president of an Asian country, which was trying to ban tobacco advertising, F1 motor racing and the importation of cigarettes.

"What's the problem, Kohl?"

"These banana republics don't know the meaning of free trade," snarled Kohl. "FloorMart provides factory employment for millions of their people and then they try to stop us selling them tobacco."

"Why doesn't FloorMart pull its factories out?"

"You kidding, Orpheus?" laughed Kohl. "We would have to make the goods in the States and that would push the unit costs up by a thousand per cent."

"But they would still cost the same in our stores."

"Naturally. But FloorMart would make much lower profits."

"That would never do. But why don't we employ all those people we have in the Under Class Areas?"

Kohl Rabi's eyebrows shot up. "You turning commie, Orpheus?"

"Of course not. I just thought …"

"Well don't think. It's very unhealthy for a president to think."

"I was thinking …"

"Careful, Orpheus."

"I was thinking that the MV Couch is in San Diego for most of the year and is not paying any taxes."

"That's because it's a charitable foundation and non-profit making."

"Yes. But what about the Bank without Frontiers? It's a commercial bank turning over billions every day. Surely we could tax the bank."

"Leave it, Orpheus. I told you it was dangerous to think. I happen to have an account in BwF. It doesn't pay interest but the money is safe and no questions asked."

Orpheus was glad he had his tape recorder switched on. That statement could prove useful at some future date. He decided to leave Brother Heinrich alone for the moment and turn his attention to Slam Duncan, the traitor.

The Eternal City was looking its best in the spring sunshine. President Quagmire immediately went to see Gloria Inexcelsis Deo.

"Don't sit in that chair, Orpheus," she said, pointing to a comfortable armchair to her left. "That huge, heavy chandelier over it might fall and kill you."

"Why don't you get the chandelier fixed?"

"It is fixed. I have a red button under my desk,

which will trigger it and thrust the unfortunate occupant of the armchair into the fires of hell. I had several just like it in the Malacanang Palace."

Living in the ex-Vatican was evidently having an effect on Gloria. She rang the bell and Luigi Parmesano came in, accompanied by a retired McDuff's Delivery Dragon.

"Allow me to introduce my Spiritual Advisor," said Gloria. "Luigi Parmesano, President Orpheus Quagmire. Don't pat the monitor lizard, Mr President."

Luigi was wearing his usual dark shades. If Orpheus had been able to see his eyes, he would have run screaming from the room.

"Do you know where that rodent, General von Rouse, is?" he murmured.

"He's called Brother Heinrich now," said Orpheus.

"Is he a man of the cloth then?"

"I'm not sure."

"No matter. If he's a real monk, the pleasure of killing him will be all the greater."

President Quagmire shuddered. He had heard the story about Luigi selling his soul to the Devil.

"So where's he living?" asked Luigi.

"On a ship, based in San Diego: the MV Couch."

Luigi made a note on his palmtop.

"He's history," he murmured

Orpheus smiled. It looked as though he was going to kill two birds with one stone: the boot for Slam Duncan and a cement suit for Heinrich. It had been a very productive meeting. He stood up and smiled down at the Gloria, the world's most powerful midget.

"I'm so grateful for your help, Gloria," he said with a beaming smile.

Gloria didn't answer and didn't smile.

Chapter Twenty-Six

Richard Trichet was preparing lunch for the crew. Being in the galley of a submersible meant that he preferred to make a majority of dishes that required no heat. He had a fantastic range of salads based on fresh produce when available, and using longer lasting vegetables when away from port. The Nemo had been at sea for two weeks, so the salads were grated celeriac and five-bean chili. There was a dish of eggs mayonnaise. Richard was proud of his homemade mayonnaise; none of your bottled stuff for him. There was a hot dish of steamed Marlin with sweet and sour sauce and steamed rice. Richard took great pleasure in giving the crew the fresh fish, because he had caught it that very morning from the deck of the Nemo.

Cooking always gave him time to think and he was remembering the traumatic months before he had joined the Nemo.

Luigi Parmesano had arrived at his secret laboratory in the Sicilian hinterland to find Dr Trickie in a Wizard's fancy dress, tall hat, stars, moon and all. Lizards, bats and a small crocodile were suspended from the ceiling The work of the lab was focused mainly on tweaking steroids to produce undetectable, performance-enhancing drugs for athletes, but the main income of the family now came from people smuggling and arms.

"Good morning, Trickie," he said. "Or should I call you Merlin?"

"Whatever."

"Look at me, when I'm talking to you."

"I am looking at you."

Luigi knew that Trickie's divergent eyes always gave the impression that he was looking at someone standing next to you. Now he was nervous, his eyes were completely out of control, each going their different way.

When Trickie started working for Luigi (an offer he couldn't refuse) he had joked, "I think I'll have a go at finding the philosopher's stone."

"Why don't you?" said Luigi. "If a nut and a radish can invent an anti-gravity motor, changing base metals to gold should be child's play."

Ever since then, Trickie had been forced to go through the motions of practising alchemy, and it was driving him mad. The trouble was that Luigi really believed the task was possible, and became very impatient when no results were forthcoming. When Luigi got impatient, he was very dangerous. Trickie could see the writing on the wall. It said, Run! So he ran.

Unable to leave Palermo openly, he joined a party of Afghans who were being smuggled to Brazil. He was herded into a twenty-foot container with twenty refugees and swung on board a ship. During the voyage, the passengers suffered from hunger and exposure since they were given very little food and the container was their only accommodation. Trickie swore to travel business class the next time. That would have meant extra pasta and a hammock below decks. When the ship reached the coast of what turned out to be French Guiana, the refugees were pushed overboard and had to swim to the shore, where the French Gendarmes guarding the Kourou Space Centre arrested them. Trickie used some of the diamonds he had made in his lab to bribe a guard to release him. He went straight to the airport and checked into an Anti-grav hotel that was leaving for San Diego. He knew that his first priority was to change his appearance.

When Trickie boarded the MV Couch, he was taken to see the Mother Superior. Apart from psychiatrists, the ship also had surgeons who performed wonders with deformed faces. Their main work was with the poor, repairing cleft palates and similar congenital disfigurements. They were not averse, however, to taking rich patients who paid huge fees, which subsidized the important work. Trickie explained that his wandering eyes were ruining his life. His nose was an embarrassment as it was twice the size it should be. His ears stuck out like elephantine flaps and he had no chin to speak of. As for his snaggle-teeth, they stuck out at a dozen different angles. He was checked into the surgical ward and transformed. Surgery gave him straight eyes and people were able to appreciate how beautiful and blue they were. He had a successful nose job and his ears were pinned back. A bone graft gave him a strong chin, and a dentist replaced his teeth with titanium implants to which were attached a beautiful set of even, white teeth. The barber cut, dyed and styled his hair; the beautician gave him a facial and a manicure. Never an impressive figure, Trickie had neglected to exercise and was rather puny. Being thin, he had a good foundation for building his body. A course of designer steroids and a high protein diet set him up to pump iron.

He was on the ship for six months, and when he was free, he indulged his hobby, which was cooking. He was in fact a very good cook, having taken a Cordon Bleu course in Paris in his younger days. His help in the galley was much appreciated by the cooks, who were pleased to learn from someone who had trained in Paris. Having handed over the last of his diamonds in payment of his bill, he was ready to leave the ship. The Mother Superior had enjoyed her Pygmalion role and was very proud of the handsome man who sat in front of her desk.

"What will you do now Monsieur Trichet?" she asked.

"I'll have to find work. I've thought of going into the restaurant trade. I am qualified, of course, but I have very little experience."

"Don't worry about that, Richard," said the MS. "I'll give you a reference stating that you have worked in our galley for the last six months. Brother Heinrich's son needs a cook for his submersible. I'll have a word with his father."

Brother Heinrich was unaware that Monsieur Trichet was in fact Dr Richard Trickie. If he had known he would never have recommended him for the job; he would have taken him back into his bank. As it was, Monsieur Trichet, transformed by plastic and cosmetic surgery, joined Frank on the Nemo. The Mother Superior immediately went into retreat for a week to do penance for the impure thoughts she had had about the handsome French cook.

Captain Hiroshi Nagasaki set a course for the Sunda Trench, south of Indonesia. Frank was particularly interested in the Indonesian Colony because of the fauna on land as well as the denizens of the deep. For example, he had never seen a Komodo Dragon, and every self-respecting zoologist made at least one expedition to find it and make a documentary to pay the bill. When the Nemo arrived in Denpassar, Bali, Frank and Orville went to see the American Governor of the island. He was a rough old boy from Texas, who knocked the stuffing out of everybody with his backslapping. He was, however, very helpful with permits and visas. Visitors were restricted because of the frequent bombings and riots, which racked the colony. Acquired in the time of President Dubya, in a pre-emptive war against possible terrorists with forty-five minute WMDs, which were never found, the colony had not known a day's peace since. The occupying forces were shocked when the population did not welcome them, in fact, it was downright resentful and ungrateful. Resistance movements sprang up and the occupying forces had to contend with losses that were kept from the media.

The small island of Komodo was a hive of activity when the Nemo arrived in port. There were twenty television crews making documentaries, and a large force of counter-insurgency troops trying to find the Kalahari Bushmen's Expeditionary Force, which was aiding the Indonesians in their fight for independence. It was an unequal struggle because the Bushmen were four feet six tall and the elephant grass was six feet tall, rendering them impossible to find. As a result, the economy suffered and was made worse by McDuff's withdrawing from the island where they had been running an experimental scheme to breed the giant monitor lizards. These were to have been trained as Delivery Dragons, but they had failed a crucial test. They delivered hamburgers all right and never ate them; but they did eat any animal, dog, cat, rabbit or pony, that crossed their path. We went into the bush to find our dragon. Rabbit advanced in a series of huge leaps over the tops of trees and from tree to tree.

"I know the lizards are partial to rabbit," she said, "so I'm going to spend as much time as possible off the ground."

Arriving at a clearing, we set about filming a monstrous beast, all of twelve feet long, which was devouring a water buffalo. A hissing sound drew our attention to the bush. When we investigated, we found ourselves looking down at a diminutive family.

"Hi!" said the Bushman, "I'm Major Dan, this is my wife, Major Barbara (her nom de guerre), and this is our son Lieutenant Bart. We are fighting a war of liberation. What are you doing?"

"We're making a documentary," I said.

Dan looked at me in surprise. It was unusual for him to meet someone smaller than himself. "Are you and the nut out of MonCul?"

"Yes, we are," I replied

"We've had GM up to here," he said, which wasn't very high. "We had a war with the Kalahari Yams, who resented the murder of President Sam W. Yam. The outcome was that we had to sign a peace treaty giving them half the country as a reservation. Not having enough land left to feed ourselves, we drifted into Windhoek; Bart got in with a bad crowd, I started drinking and our family went to pot. Finally we pulled ourselves together and joined the World Revolutionary Party; and now we fight in Liberation Movements."

"Did you meet Sam W. Yam then?" said Orville.

"We did, and he was delicious. But little did we know that we were eating ourselves into a disaster."

At that point a platoon of Marines came crashing through the bush and the little people melted away.

"Which way did they go?" yelled a Sergeant.

"Thataway," bellowed Coco, pointing to the sea.

Having satisfied Frank's desire to meet a dragon, Captain Hiroshi took the Nemo to the Sunda Trench. There we were to meet a fabled creature.

Chapter Twenty-Seven

D'Arcy Versey, Conchita and little Astair, Fred for short, were walking along the Panama seafront. D'Arcy had suffered a financial setback in Los Angeles, where an investor wanted his money back. He has taken a share in the remake of the Sound of Music, which was never remade. D'Arcy turned out to be hopeless in the Captain's role and Maria, newly released from the Couch, had relapsed into a catatonic trance. A quick exit was in order and Panama was a logical destination; after all, he had once been President of the country and people still remembered his musicals with affection. The current president was Leonardo Parmesano, grandson of Luigi. He proved to be a very efficient organizer and placed his soldiers in every position of importance. Panama was a refuge for the world's miscreants and the only rule was against robbing or murdering anybody inside the territory; to do that the victim had to be taken out of the country. There was no crime in Panama.

"This country is unbelievable," said Conchita. "It's had more presidents than I've had hot chili peppers. Why can't they leave the government alone?"

"It's in for a period of stability," said her

husband. "The Americans have taken the Canal back in a deal with Leonardo. His appointees control Central Snoopers. When he learns to read and write, Anthony Capote, the Secretary for Education, is tipped to succeed Leonardo. We have to get an interview with one of them if we are going to restore our fortunes."

Anthony Capote, a Parmesano appointee, was reading the riot act to the Vice-Chancellor of the University of Panama, as D'Arcy waited in the anteroom.

"What for you no give me Doctorate?" he shouted. "I secretary for Education and I ain't got no degree nor nothing. I wanna Doctorate."

"Well, you can't have one, so there," said the Vice-Chancellor. "You have to be able to read and write to get a Doctorate."

"Then you teach," said Anthony.

"Fine," said the Chancellor, glad of a way out, "I'll put you down for the Adult Literacy Programme."

"That easy?"

"Very easy. Any stupid moron can learn to read on that programme."

The Chancellor rushed past D'Arcy and out of the building. Back in his office, he started filling out applications for posts outside Panama.

D'Arcy entered.

"Sit!" said Anthony. "What for you wanna see me?

"I'd like to offer my services to organize the teaching of dance in the territory."

"You stoopid or something? Dancing ain't no use"

"Well, it can be," said D'Arcy. "There used to be a large number of schools which taught dance in Panama. And the Academy of Performing

Arts was world famous. Our dancers were in most of the major ballet and dance companies. A good percentage of our GDP came from the arts. Do you know, at one time we had ten companies touring Les Miserables, Cats, Singing in the Rain and The Sound of Music among others?"

"No. Can you read and write?"

"I beg your pardon?"

"Can you read and write?"

"Of course I can."

"Can you teach me?"

"Sure, I can."

"OK. Here's the deal. You teaches me read and write. I makes you Secretary of Performing Arts."

Determined not to fall down on his side of the bargain, D'Arcy insisted that Anthony should have at least two hours a day in the classroom.

He soon found out what the Secretary for Education's problem was. He made good progress with phonic words, but when he met a non-phonic word like 'through', he threw his desk out of the window and refused to continue.

"English is a moron language what don't know how words is wrote," he screamed.

D'Arcy calmed him down by playing Karaoke discs of traditional nursery rhymes and songs. Anthony had missed out on this important cognitive development stage, because he had become a Mafia soldier at the age of three. Gradually he accepted that some words were 'look and say' to be recognized by their shape. After that he progressed in leaps and bounds in both reading and writing, but his grammar failed to improve. He refused lessons on the grounds that his syntax was already perfect. D'Arcy didn't argue, but insisted on writing Anthony's speeches.

The quid pro quo exceeded all D'Arcy's expectations. He was given a completely free hand in cultural affairs provided he stayed within budget. For once in his life, he watched the cash like his life depended on it, which it did. He organized a season of Italian opera, which delighted the people who mattered. He persuaded a leading conductor to come out of retirement to rehearse the Panama symphony orchestra. The Panama ballet soared to new heights under his guidance. He also knew how to get other countries to provide freebies in the form of cultural visits. The Shakespeare Company came from the American colony of Britain. The Bolshoi Ballet came from Greater China. Folk ensembles came from a dozen countries.

One spin-off was the conversion of Anthony from an ignorant, illiterate oaf, to a sophisticated, sensitive patron of the arts, who never missed a concert or play and even started writing his autobiography with the title "Out of the Darkness'. When it was published, it was short-listed for several literary prizes. His style was praised as innovative since it ignored all the syntactical conventions. If he had been an American Commonwealth citizen he would easily have picked up the Booker Prize.

One day Anthony was reading the collected poems of Robert Browning, a long-neglected Classical poet, when his eye fell on the lines:

Too gold the light grew

Golden and not grey

He realized it was an omen. He was having far

too good a time for it to last. He needed some shade to contrast with the light. He had recognized in himself the fatal flaw of *hubris*. He was becoming proud and therefore heading for the kind of fate that befell the protagonists in Greek tragedy. He decided to do something about it.

He often watched the Brother Gerbil Golden Hour and he had even sent a donation. He found the mellifluous tones of brother Gerbil recommending the sloughing off of the burden of earthly cares and wealth, especially wealth, rather inspiring. He went to see President Leonardo Parmesano and put it to him that a retreat at Brother Gerbil's Ashram might be just what the doctor ordered. The President answered calmly, but coldly, "By all means, take a week, on one condition: at the end of the week, you kill that thieving rat, Gerbil."

Anthony considered that a fair arrangement. He made a booking at the Ashram and packed a bag and his revolver. Before he could set off on his pilgrimage, Gerbil had got wind of who was coming, and got the wind up. He told his followers that he was going into retreat in the hope of ascending to a higher plane, and he wouldn't be coming back. He gave his devotees the key to his wine cellar and told them to celebrate his prospective elevation. As good as his word, he raced to the Golden Hour Airport and ascended to the heavens in a Buddhist temple bound for Bangkok. On learning of Brother Gerbil's ascension, Anthony cancelled his booking. As an act of self-abasement, he became a vegetarian on Monday and fell to devouring a whole roast chicken on Sunday. He tried walking backwards on a pilgrimage to the Great Cabbage in Rio, but kept falling over and had to give up after half an hour: he was not the stuff of saints.

"It's no good," said Anthony to D'Arcy, "I'll just have to face the fact that I am a proud man and accept the punishment of the gods when it comes. What did Robert Browning know anyway? In the meantime, what's your latest hit show?"

"Becket's *Waiting for Godot* at the Lyric Theatre," said D'Arcy. "It's a cathartic experience, which I am sure will do you a lot of good. Afterwards, we can have supper at the Café de Paris and take in the floorshow. My Cancan Dancers are performing. They are really wonderful, although their underwear is a little worse for wear from doing the splits night after night."

Chapter Twenty-Eight

Gloria Inexcelsis Deo watched President Orpheus Quagmire leave her office. She thought about calling him back, and asking him to sit under the lethal chandelier, but he would never have fallen for such a stratagem. She picked up her phone and called the Bank without Frontiers. The idea of liquidating Brother Heinrich and perhaps jeopardizing the sanctity of the Bank filled her with horror. Quagmire was going too far, encouraging Luigi Parmesano in his murderous vendetta. Brother Heinrich may have converted Guido Parmesano into pebbledash with a tank shell, but that was no reason to kill him. Ever since the days of the Seascam, her predecessors had transferred most of the cash coming into the Philippine Treasury to their own accounts; she was no exception. Her salary was derisory and she needed lots of cash to buy apartment blocks and shoes.

"Yodel," she said, "It's so nice to talk to you. "And how is Friday?"

"Very well, Gloria," said Yodel Lay. "Your last transfer arrived safely. It was from the tobacco tax, I believe."

"That's right," said Gloria. "The Widows and Orphans money will be arriving next week. I have a very important matter to raise with you. That maniac, Luigi Parmesano is putting out a contract on Brother Heinrich. He was egged on by Orpheus Quagmire. If they hit Heinrich you are for the high jump."

"You mean we, don't you Gloria?"

"No, Yodel, I mean you."

"Oh. So what shall I do?"

"I have a video recording of Quagmire making a racist remark. I'm sending it to you and I want you to give a copy to all the news channels. After it appears, I want you to tell all the senators you have got the dirt on, which is all of them, I believe, that they are to impeach Quagmire and boot him into oblivion where he belongs."

"I shall be done, Little Lady."

"Not so much of the little, Fatso. Just remember I am the most powerful woman in the Universe."

"Of course Madam Chairman."

"And another thing, Quagmire wants to sack Slam Duncan for committing him to the funny farm and then taking his job. I'm not having it. Duncan is an excellent Head of Central Snoopers. He's helped me to put away the people who were accusing my husband of corruption."

"I thought Cox Pippin was the boss of that outfit."

"Nominally, but he's so rotten, he has practically liquefied. Duncan runs the show. Have a word with Kohl for me."

"Consider it done."

Yodel Lay put down his phone and turned to his partner, Friday.

"The Midget is getting too big for her many pairs of boots," he said. "I don't give her more than six months in that job."

He got in touch with the Secretary of State, Kohl Rabi.

"Kohl, your man Quagmire is interfering where he's not wanted. You should really keep him in order."

"Now what's he done?"

"He is trying to get rid of Slam Duncan and has told Luigi Parmesano where Brother Heinrich is, so he can be hit."

"Why would Luigi do that?"

"Heinrich splattered his son against the wall of the Panama Presidential Palace with a tank shell."

"Nasty! But Heinrich has changed and he's looking after our money."

"That's the point. Gloria is jumping up and down."

"Ah, we may catch a glimpse of her that way."

"Be serious, Kohl. You must stop Quagmire and Luigi."

The conversation between Gloria Inexcelsis Deo and Orpheus Quagmire hit the news channels. Every pressure group rose up in fury at his blatant racism. The Ku Klux Klan made things worse by coming to his defence and setting up burning cabbages. The Get Whitey Coalition started training to overthrow the government and

Friday Knight addressed a million man rally in front of the White House. Quagmire had to rush back from Rome to face impeachment proceedings. The witnesses against him were numerous. A torn black dress was displayed as evidence that he had deliberately assaulted it because of its colour. A little girl tearfully told how he had thrown her black doll into the Potomac. A lady of the night told how he refused to accompany her because she had a black grandmother. The clincher was an appearance by Slam Duncan who described in lurid detail the madness of Quagmire. He became ex-president, was immediately pardoned for any crime he might have committed, and allowed to keep most of the White House furniture to equip a mansion presented to him by a grateful nation.

Luigi Parmesano was a tougher gander to cook. Slam Duncan had him shadowed to a remote island in the Atlantic, which he often visited to pay homage to Napoleon Bonaparte who had died there. The Seventh Fleet immediately blockaded the island and Luigi was only allowed to leave after signing an undertaking not to harm Brother Heinrich and to take a course in anger management. Admiral Graf Spee did not see that Luigi's fingers were crossed when he was signing the surrender document.

Luigi went to Rome where Kohl Rabi was discussing affairs of state with Gloria Inexcelsis Deo, and people were starting to talk. He was invited to Gloria's office for a brainstorming session.

"Hello. Luigi," gushed Gloria. "It's so lovely to see you."

"I'm glad you think so," said Luigi, taking off his shades. "Hello, Kohl. You still around? People are starting to talk, you know."

"Let them talk," said Kohl, giving Gloria an adoring glance. "All publicity is good."

"Not the kind you are getting."

"Please sit down, Luigi," said Gloria. "The armchair on my left is very comfortable."

A younger Luigi would have sensed danger in the specificity of the invitation to a particular seat, but he was old and about to not get any older. He sat. Gloria felt for the red button under her desk. The brainstorming session ended with the braining of Luigi Parmesano with a two hundred pound chandelier. Luigi hadn't read the small print on his pact with the Devil, which excluded death by light fittings.

The news spread like wildfire. Mafia Dons from

all over the world converged on Rome to accompany the dear departed Don to Sicily. Gloria sent a message of condolence, but it did not fool the Parmesano family. They knew the Don's death was no accident and blood demanded blood. The Sicily-Filipino War was soon to break out. Luigi was laid to rest in the family mausoleum in Palermo. Leonardo Parmesano was summoned back from Panama to take charge of the family. The presidency of Panama was up for grabs again.

D'Arcy Versey knew that continuity was vital, both in the arts and in life.

"Just when we get things going nicely," he said to Conchita, "a fly has to drop into the ointment."

"I don't see the problem," said Conchita. "Anthony Capote will do a good job now he is president. For one thing, your Arts programmes will get his support."

"I'm not sure about that. I remember the old Anthony. He could revert to type at any moment. His culture is simply a veneer hiding a crude interior. I heard him talking about Descartes to the Professor of Philosophy at the University. He gave a most impressive summary of the *Discours*. I happened to see his copy of the book. Anthony had reproduced the blurb word for word. That's what he does; he reads the blurb on hundreds of books. He has an amazing memory and can regurgitate them."

"His English is much improved."

"True. I found out that he had watched My Fair Lady many times when he answered one of my funding requests with 'Not bloody likely'. He knows the dialogue by heart and uses a good deal of it in conversation."

Much to D'Arcy's relief, Anthony did not revert to type. The success of *Out of the Darkness* inspired him to continue writing. The book became bedside reading for miscreants with a bad conscience. Anthony continued in the same didactic vein in his next book, *It's never too late,* which led to a huge drop in the consumption of alcohol, tobacco, gambling and the crime rate. This went down very badly with the Mafia. Their casinos started losing money; businesses, offering special services for tourists, were closed down by their conscience-stricken owners. Charitable donations soared, benefiting from the cash that wasn't being gambled away.

Responding to an urgent summons from the Panamanian family, Don Leonardo Parmesano arrived in the Sistine Chapel, which he had bought from the Seven Day Vegetarian Church and fitted with Anti-gravs.

"Anthony, Anthony," he said as he gazed on the wonderful ceiling. "What do you think you're doing? Where will our businesses be if people stop smoking and we can't smuggle cigarettes? What will happen to our alcoholic beverages business if people stop drinking? What will happen to our casinos if people stop gambling?"

"Everybody will be healthier and better off," said Anthony. "There'll still be a market for wine. I drink a sip at Mass every day."

Leonardo couldn't believe his ears.

"What? You've gone back to the old religion?"

"Many people are rediscovering the old faith," said Anthony. "We've reclaimed the Panama City Cathedral, which the Seven Day Vegetarian Church stole from us. Next year, the Vatican; the year after, Jerusalem! Millions are joining us in a Crusade for fish on Fridays, and meat every other day. The story that Jesus turned water into Coca Cola at the wedding in Canaan is nonsense put about by the Vegetarians. It was good, honest, fermented grape-juice; wine in other words."

"I don't doubt it," said Leonardo, "but you're ruining our businesses."

"You still make millions from people smuggling."

"The bottom has fallen out of that too," said Leonardo. "We have smuggled so many people from Central Europe, Asia, Central and South America and Africa to Western Europe and North America that North Americans and Europeans are flooding into the depopulated regions and setting up ranches and plantations with huge tracts of land. There's nobody left to smuggle!"

Anthony was thoughtful. "I'll tell you what I'll do. My next book is entitled *The Golden Mean.* It will recommend moderation in all things; in other words: the odd glass of wine, the occasional cigar and a little flutter on the horses. I can't say fairer than that."

Leonardo took off his shades, revealing that he had inherited his grandfather's eyes.

"It's a lovely day, Anthony," he said. "Why don't we take the Sistine Chapel for a spin over the ocean?"

A few miles away, D'Arcy Versey felt a cold shiver run down his spine. "Somebody just walked over my grave," he said to Conchita.

Chapter Twenty-Nine

Hiroshi Nagasaki turned to Frank and Orville.

"Here we are," he said, "over the Challenger Deep: all 35,838 feet of it. You could fit Everest in it with a lot to spare."

"Thanks, Hiroshi," said Orville, "I suggest we spend some time on the surface, have a picnic on deck, and take in a glorious sunset. Our next dive will be a long one, as I want to film and take samples of everything we see."

Rick, our fantastic cook, excelled himself with a feast worthy of a five star Michelin restaurant. The food eaters started with live oysters, which had been growing in a wire basket, suspended over the side. A Bouillabaisse worthy of the best Marseillaise restaurant, with garlic croutons floating on the surface, was followed by Swordfish steaks, from a fish caught that very morning by Rick, served on a bed of asparagus (frozen - but you can't have everything at sea). There was freshly baked bread throughout, and a tarte aux pommes with thick cream, and a cheese platter rounded off the feast. Rabbit asked for a

Waldorf salad, but Rick didn't have any lettuce. Instead he served her an excellent salad of carrots, turnip and croutons. We all relaxed as the ocean lapped gently against the side of the Nemo, and the sun slid below the horizon in a blaze of reds. We slept well, for the next day we were going to start our descent into the deepest trench in all the oceans: the Challenger Deep.

The descent lasted three days. Below a thousand feet, we spotted very few species, but the ones we did see became more and more bizarre, unlike anything we were used to finding in the upper levels. We filmed and snared as many of them as we could. Capturing specimens was difficult because the enormous water pressure was affecting the hydraulics of the remotely operated gear. At thirty thousand feet, the Nemo suddenly shook and then swayed from side to side. From the bow observation ports we could see enormous tentacles floodlit by our powerful lamps; a huge mass had us in its embrace.

"It's a Giant Squid," said Hiroshi calmly. "The one described in legend isn't so far-fetched after all. And from the way we are being pushed around, I guess it is a male that thinks the Nemo is a female."

We were all alarmed and afraid that our capture gear would be torn off. Any damage to the craft at this depth and it would implode. We had a system devised for repelling inquisitive sharks. Hiroshi switched it on and put a burst of high voltage electricity through the outer skin of the Nemo; we were tossed aside like a spurned lady squid. We filmed the monster as it jetted away from us. It was almost as long and as broad as our sub. We glimpsed its beak, which looked capable of opening us up like a tin can.

Undeterred, Frank ordered Hiroshi to take us to the bottom where there were many strange species. We filmed and collected samples, without a hitch. Apparently the Giant Squid hadn't done any damage. If it had, we wouldn't be around to tell the tale. We spent two days on the bottom and then started the ascent.

"I would have liked to spend longer on the bottom," said Frank, " but we had better not push our luck."

We did not see the Giant Squid again but we did meet our friend, Whale, who accompanied us to the surface, chatting with Coco and me all the way up. Rabbit showed his limitations when he confessed he was unable to communicate with Whale since he had not been programmed for sea mammals.

"I don't suppose Orville expected me to go to sea," he said. "I'll ask him for the programme when we get home."

Whale gave us a complete run down, and run down were the operative words, of the state of the oceans, which we passed on to Frank and Orville. There was a glimmer of hope. The worldwide vegetarian movement directed by the Seven Day Vegetarian Church had helped in reducing the numbers of fish caught. The WO had banned the catching of some species, but, as usual, the Scandinavians and Japanese hadn't taken much notice. They had even killed a few whales 'for research'. Whale was very indignant about that and asked us to write a protest letter to Emperor Sashimi when we got back. There was a cloud on the horizon in the form of the resurgent old religion, which was bringing back the refrain: fish on Fridays, meat every other day. When we surfaced, Frank, Orville, Rick and Rabbit all emerged to breath the invigorating sea air. Coco and I smiled indulgently on the weakness of the air breathers.

We had been out of radio touch for a week, so Coco put the latest world news on his speakers: Mars had been invaded again and the biological WMD defeated by a new vaccine, which had protected the invading force. Every last Martian had been killed, although no bodies were found, because Martians vaporized as soon as they took their last breath.

War between Sicily and the Philippines had broken out. The Sicilians had detained fifteen thousand Filipino Domestic Helpers in concentration camps, in an attempt to force the Philippine Government to surrender.

The Sistine Chapel had returned to Palermo from Panama with only Don Leonardo Parmesano on board. He denied that the

President of Panama, Anthony Capote, had been on board when the Chapel left Panama City.

In sports news, the New York Yankees had been swapped for a panda, with the government of Greater China, and a rabbit had broken the world high jump record.

"That wasn't me," said Rabbit. "It was the clone that Orville gave to the Delightful Dim Sums."

We were tempted to return to the bottom of the Challenger Deep after listening to that baloney. Instead, we sailed north-west to the Philippine Trench.

We surfaced in Manila Bay, emerging through a blanket of detritus. We tied up at a jetty and were immediately besieged by a crowd demanding proof that we weren't Sicilian spies. Numerous bribes were paid to officials, who threatened to kidnap us if we didn't pay up. On the spot ransom was in favour because it avoided the expense of board and lodging for the victims. The capital was denuded of troops, who were all in the south, fighting the rebels for a share of the millions of dollars in ransom that they had just received for a group of missionaries. So when a fleet of gondolas sailed into Manila Bay, a crowd of admiring females gathered to listen to the Gondoliers singing excerpts from *The Gondoliers*. The handsome rowers came ashore to be mobbed by their admirers. Taking advantage of this diversion, a small army of mountain shepherds, some with sheep, rose from the bottom of the gondolas and came ashore. Armed with AK47s, disguised as shepherd's crooks, they made their way to the Malacanang Palace.

Charlie Bananaboat Deo Deo was looking after

the shop while his wife, Gloria, was running the world from Rome. The shepherds came up the drive and rang the bell. Charlie answered the door himself, as all the servants had left because their wages hadn't been paid. He was furious with his wife because she had emptied the treasury and hadn't given him any housekeeping money.

"Go away," he said. "I don't need any sheep today."

Garry Baldi, the leader, shoved his loaded crook under Charlie's nose. "We're coming in," he said.

The palace filled with Sicilian shepherds, who proceeded to slaughter a sheep and roast it in the middle of the grand ballroom. Baldi took Charlie into the presidential office and sat him under a chandelier. He found the row of red buttons under the desk.

"I know about these lights," he said. "One wrong answer and your lights go out. I have here a document of unconditional surrender, allowing Sicily to annex your country. Sign it."

Charlie signed it. He looked at Baldi. "Could you find your way," he said, "to making a token down payment for the country. I haven't eaten for days."

"Help yourself to some roast mutton," said Baldi

"You do know," said Charlie, "that the country is supposed to be independent but it is really an American colony. That document is invalid. The Americans and my wife are going to be furious."

"What do we care?" said Baldi. "We own the country now."

In Rome, Gloria was talking to Kohl Rabi about the annexation of her country by Sicily.

"What do you intend to do about it?" she said.

"Have you got a new president, yet?"

"The Vice-president is in charge," said Kohl.

"But he's Pot of Yogurt!" said Gloria.

"He's doing a good job," said Kohl. "That is, he's doing precisely nothing, which, in the circumstances, is the right thing to do."

Kohl had been briefed on the situation in Sicily.

"My dear, little cabbage," he said.

"Not so much of the little," said Gloria.

"Sorry," said Kohl penitently. "The women of Sicily have risen against the government. Their domestic helpers were herded into concentration camps when the war started. This meant the Sicilian women had to do things, which they had never done before: take care of children, do housework, cook, walk the dog and wash the car. There are no nurses in the hospitals, and patients are having to empty their own bedpans. There are no chauffeurs to drive the kids to school. It's a mess. The latest intelligence is that the women have withdrawn all marital privileges and have brought the government to its knees. It's only a matter of time before the internees are released and a peace treaty is signed."

When the news of the Sicilian government's surrender reached Manila, the Gondoliers set up their own transport company and started plying between the islands in their gondolas, which were fitted with Ant-gravs. (How do you think they got from Venice in the first place?) With all the old ferries put out of business, the number of disasters fell from 75 a year to zero. The number of passengers drowned fell from 1,540 to three drunken sailors. The shepherds decided to stay on, married local girls and started sheep farming in the mountains. The war was over.

Frank, Orville, Hiroshi and Rick decided to take some shore leave. I went with them; Rabbit and Coco stayed on board to make sure no intruders got into the sub. We already knew that anything not nailed down would disappear. Well, we could hardly nail the sub down. We had just checked into a hotel, when I got a call from Coco saying that a gang of thieves was towing the Nemo to a yard, where they proposed to sell it for scrap. Orville told him to take evasive action and lose the subnappers. Coco was well capable of handling the sub. Rabbit jumped onto the console and, following Coco's instructions, punched all the right buttons, causing the Nemo to submerge, dragging the pirate boat under. It then surfaced, Coco untied the line and, together with Rabbit, guided our craft smoothly back to port. By this time, Hiroshi had arrived and took command. He decided to stay on board in case of further trouble.

We were well acquainted with the Under Class Areas in the States. They were small pockets where undesirables, like communist agitators, used car salesmen and real estate agents were segregated from the general public. The system in Manila was the reverse: the super-rich were segregated in luxurious ghettos, while the rest of the city was one huge Under Class Area. Our hotel was in the Central Business District, walled off from the rest of the city. We were told to be back in the enclave before sunset, because the gates were locked and not opened until dawn the next day. We didn't want to go out after dark anyway, because a light shower caused the streets to flood, bringing out gangs of young men and boys to push cars that had stalled in the deeper puddles. For them, it was the only paid work they knew and they made sure that the street drains were permanently blocked with rubbish, and all the cast iron manhole covers stolen and sold for scrap. There was a rising tariff for a simple push, a lift out of a manhole and push, and a lift out of a manhole and push to a service station for major repairs. The gangs handed this out on printed cards.

From what we saw, nothing was ever done in the city. When the World Bank coughed up money for a development project, there was a feeding frenzy in the government and the money disappeared. It was the only place I had seen where there were roads that led to nowhere. The official policy seemed to be that if it was any use it would require costly maintenance, therefore it was better to build something and not use it. The fiction of independence was maintained, but the fact was that it was a tightly controlled colony where dissent was jumped on by the colonial masters and the priests of the old religion. They told their congregations exactly what to do, what not to do and who to vote for in the rigged elections. The faithful would then get their reward in heaven. We were all glad to go back on board the Nemo and put out to sea.

Chapter Thirty

Panama was in turmoil. No trace of President Anthony Capote had been found and he was presumed either dead or bewildered and befuddled. The Vice-president was a Packet of Spaghetti, chosen for his patriotic symbolism. Only when the Secretary for the Performing Arts went to see him, did his staff realize that he had been served up with Bolognaise sauce by a careless cook. D'Arcy was distraught. With

Anthony and the Vice-president gone, he was not protected from the slings and arrows of his jealous colleagues. Lesser actors, musicians, dancers and directors started to savage D'Arcy Versey's work, calling it old-fashioned, out of date and irrelevant in the modern arts scene. They tried to install a ballerina's left shoe as Secretary for the Performing Arts, and only failed because it was found to be the wrong size for the job.

Help came in the form of the MV Couch. The looting and destruction that followed the disappearance of the President caused massive food shortages, and wanton acts of violence traumatized hundreds of Panamanians. The Sisters of Commiseration started work as soon as the boat docked. The crew offloaded tents for temporary shelter. Food was distributed to the most needy. Brother Heinrich and Yodel Lay persuaded the government Ministers to stop looting their own houses and hold a presidential election. They then bribed the Ministers to agree on the candidates: a Panama Hat and D'Arcy Versey. Hat won the election and D'Arcy became Vice-president. Whenever he appeared at official functions, D'Arcy wore Panama Hat, and wore out the joke that he was wearing his presidential hat.

I was following the events in Panama and suggested that we pay a courtesy call on our friends. Frank could see his father, and we could all enjoy the new Vice-president's theatricals. Frank was reluctant, but Orville and Rick were keen to take a decent shore leave and see some old friends. We arrived in Panama, went on board the MV Couch and found the wards full of trauma cases. The Sisters, Doctors and Nurses were all busy with their duties, but there was no sign of the Mother Superior. Then I noticed a lady in civilian dress, nursing a patient: it was the MS, who had apparently left the Order. Gone were the voluminous black habit and snowy wimple; she was revealed as an attractive woman of about forty, with blond curls and a very pleasing figure. Brother Heinrich told his son that the MS had realized that she had lost her vocation and could no longer be a nun. I saw her reaction when Rick came into the ward. She stared at him and blushed to the roots of her beautiful blond curls. In spite of the transformation, Rick recognized her at once. He moved to her as on in slow motion and took her hand.

"Guinevere," he murmured.

"Rick," she whispered.

"In the midst of suffering, disasters and war, the human spirit always shines through," said Heinrich. "But the union of Guinevere and Rick will have to wait until this lot is sorted out."

"It looks as though we have lost our cook," said Orville. "What a shame. We'll never find another like him."

Rick joined the MV Couch as a cook and he and Guinevere tied the knot six months later.

I saw that Brother Heinrich was staring at our cook.

"Is it? Can it be? Is it possible?" he stammered. "That voice. It's Dr Dickie Trickie."

Winded by the shock of his discovery, Brother Heinrich went to his cabin to lie down.

Our next call was on the Presidential Palace, where D'Arcy was holding court for a group of visiting throat singers from Outer Mongolia.

Although he couldn't stand the noise they made, which reminded him of a prolonged burp, he welcomed the group and gave them an invitation to sing at a concert one evening when he would be busy rehearsing his next musical. When we arrived, D'Arcy asked the group to demonstrate their art for our benefit, which they did. Rabbit, who had very sensitive hearing, showed his lack of sophistication by tying his long ears in a knot. We all applauded politely when they had finished burping. After they had gone, we were all silent.

"I agree that throat singing is an acquired taste," said D'Arcy, "but the President of their country has agreed to take a touring production of *White Horse Inn* which should make a packet for me, us."

He took us to his living quarters to meet Conchita and young Astair, Fred for short. The boy entertained us with a dance routine, which ranged from tap to ballet. Rabbit, the eternal show-off joined him in a Flamenco number and demonstrated his ability to thump the floor at twenty thumps a second.

Frank, who wasn't at ease in company, said, "So what do you want to be when you grow up, Astair?"

Astair, Fred for short, looked at him derisively. "Well I don't want to be a grownup, who asks stupid questions."

Even Frank laughed along with us. "Quite right," he said. "Just testing."

D'Arcy gave us a briefing on Panamanian affairs. The Colonial Power had forced him to allow troops to be stationed in the Canal Zone.

"According to the treaty, by which it returned to the US, the canal is supposed to be a DMZ. I'm afraid General Quagmire is in the Zone and, at the first sign of democracy, he will bomb us and send his tanks in. Knowing him, it will be vice-versa."

Orville and Frank looked puzzled. "Quagmire?" they asked in unison.

"Yes," I piped up. "He's back with a vengeance. He threatened to tell the media that Kohl Rabi was having an affair with Gloria Inexcelsis Deo, and neglecting affairs of state. Kohl had to give him a command, so he sent him to the Canal Zone."

The mention of certain key words evidently triggered a response in Central Snoopers, and a Black Death helicopter appeared at the window of D'Arcy's quarters. It hovered silently, thanks to the Anti-gravs with which it was fitted, and focused its camera and listening apparatus on the room.

"That's typical," said Coco. "We invent a motor to stop pollution and it's misused by the government." He pressed a button on his shell and a drawer slid out. "I have a jamming device here," he said. "Central Snoopers are now watching a home video of General Quagmire broiling steaks on his barbecue. That will get tongues wagging. Where could a General find the money to buy that quantity of beef?"

We said farewell to our friends and went back to the Nemo. Hiroshi was on the dock, looking upset.

"The Nemo has been impounded," he said, "on a charge of spying for the Kalahari Expeditionary Force."

"What nonsense," said Orville. "Who told you that?"

"That man in the scuba gear," said Hiroshi.

We went over to the diver, who was examining the footage he had videoed under the Nemo.

"I'm Lieutenant Bends," he said. "All that equipment under your craft is strictly illegal. You have apparatus for trapping fish. Fishing without a permit is a serious offence. The cameras are unlicensed in Panama and could be used for filming our anti-submarine defences. That is attempted subversion, a capital offence."

"What's all this about spying for the Kalahari Expeditionary Force?" said Orville.

"You assisted the flight of insurgent Bushmen on the island of Komodo," said the diver, "where you were photographed by a Marine."

"Just a moment," said Orville. "You're not Panamanian."

"Certainly not," said Lieutenant Bends. "I'm US Navy to the core. I swam over from the Canal Zone to arrest your spy ship."

"Well you can swim back," said Orville, pushing him off the dock. "And if I catch you here again, I'll tie a knot in your breathing apparatus."

General Orpheus Quagmire, Commanding Officer of the Panama Canal Zone, was incandescent with rage.

"How did D'Arcy get hold of that barbecue footage?"

Major Pettit, his four foot eleven second-in-command didn't know. Since his betrayal by Slam Duncan, Quagmire had avoided having assistants over five feet tall. He used to say, 'Let me have small men about me. Yonder beanpole has a lean and overbearing look.' Quagmire got onto Central Snoopers and spoke to Cyril Liberty, the local chief spy.

"I've looked into the matter further," said Cyril and it appears there were two other people,

Francis Frank and Orville Spandau, with D'Arcy, also a nut, a radish and a rabbit. The nut is the likely source of the video. He has more communications systems than you've got brain cells."

"What's that supposed to mean, Liberty?"

"I mean the nut could have jammed our Black Death spy copter."

"And where did these spies come from?"

"They're on a marine survey submersible. The Navy tried to impound it for us, but the boarding party was repelled."

"They give up too easily," said Quagmire. "I'll deal with this matter personally. Is there anything else I should know?"

"That commie boat, the Couch, is still in port, giving the Panamanians ideas about democracy and all that nonsense."

"I'll look into the boat, too. By the way, any

news on who is the next man for the White House?"

"It's a strong field, this time. There's Sammy 'Stinks' Onions, the popular stand-up comedian, Stuart Stewart, the queen of the television homemakers and a greengrocer's shop of vegetables, all fighting for the Green Party nomination. But Stinks Onions is the front runner."

General Quagmire adjusted the rear view mirrors attached to his glasses and caught Major Pettit about to put on a pair of shoes with six-inch platforms.

"Wear those, Major," he said menacingly," and it's back to sweeping out the inside of ICBMs for you."

"Sorry, General," said Pettit, "but I was going to the canteen for lunch and the waitresses keep patting me on the head as they pass me. It's most humiliating."

"Pettit, you really are stupid if you think a few extra inches are going impress the waitresses. Just watch your step."

Once Pettit had gone, Quagmire uncovered a map of Panama, marked 'For your eyes only' and wondered once again what was wrong with that particular security classification. The map was his plan for the invasion of Panama, to be implemented prior to the arrest of all the undesirables such as actors, singers, dancers and musicians, who were turning the state into a hotbed of subversion. He called the Commander of the midget submarine, Hugo Untersea, to his office. Hugo was a real midget and his best friend.

"Hugo, it's so nice to see you. How are Minnie and the little ones?"

"Thriving, Orpheus," said Hugo. What can I do for you? "

"You and I are going on a mission, and I'll need your midget sub."

"Say no more, Orpheus."

Quagmire said no more.

Chapter Thirty-One

President 'Stinks' Onions was touring the White House. He arrived at the speechwriters' office.

"You guys," he said, "had better sharpen your pencils and get into funny mode, because I'm going to need lots and lots of jokes. What's your name, old man?"

"Professor Bar Chart," replied the retired Harvard professor.

"Say something funny."

"I am an economics expert. I keep this country on the right track and balance the budget."

"That is funny," said Stinks. "Keep it up. And what do you do, little lady? "

"I'm Farah Share, the presidential advisor on women's affairs. I try to make sure we get a fair deal when it comes to employment, education, Medicare and so on."

"Good luck with the 'so on'. How about doing lunch tomorrow?"

"I'd love to, Mr President."

"Dutch treat, of course."

"Of course, Mr President."

"And you, my boy. A bit young to be a speechwriter, aren't you?"

The sixteen-year-old boy Stinks was addressing stopped reading his comic book and gave the president his full attention. "Pardon?' he said.

"What do you do here?"

"I'm Harvey Hacker. I sharpen pencils and fix the computers when they go wrong. I show the staff how to hack into games sites and other people's files."

"Isn't that illegal?"

"Only if you are found out. It's very interesting. Only this morning, I intercepted an e-mail from Kohl Rabi to the Chairman of the WO, asking her to marry him."

"And they're both married already. Listen up all of you speechwriters. I want twenty jokes about the Kohl's romance by nine tomorrow. Now get to it."

With that, Sammy 'Stinks' Onions walked out of the room. His next call was on the Secretary of State.

"Kohl, my cabbage," he said, "what have you been up to?"

"I don't know what you mean."

"Inexcelsis Deo?"

"I don't know what you are talking about."

"Come off it, Kohl," said Stinks, "every other hacker in the world got into your email and then told the other half about it."

"I shall deny it. I'll say it's a plot to discredit me and my beautiful Gloria."

"So you're going to renounce her, and tell the Senate Committee 'I never had an affair with that woman'?"

Kohl was silent for a while. "Thanks for telling me, Stinks. The main protagonists are usually the last to know when something like this gets out. I'll let you have my resignation this afternoon."

"And what reason will you give? It can hardly be 'to spend more time with my family'."

"I'll tell the truth."

"That will make a change, Kohl."

Gloria caught Kohl's full confession on the

news, and in five minutes a mob of television reporters was streaming into Parmesano Square (formerly St Peter's), demanding a statement from the Chairman of the WO, who was already screaming at Kohl Rabi on her cell phone. Nothing being easier to eves-drop on, this exchange went out to millions of amused citizens.

Gloria: Idiot!

Kohl: My Little Cabbage?

Gloria: I'm not little and I'm certainly not your cabbage,

Kohl: We have to face the world.

Gloria: You can please yourself, but I've had it with you. Why couldn't you lie, like everybody else? Half your government is carrying on with their Interns and the other half would like to but is too old. The President has a wife in Washington, and a girl-friend in just about every other city in America. Talk about a stand-up comic, he's more of a lie-down comedian.

(Television crews started to converge on the White House.)

The Secretary for Commerce has a boyfriend in Seattle. I could go on.

Kohl: Don't, my love. You never know, someone might be listening.

(A roar of laughter rose above Rome.)

Did you hear that?

Gloria: It's probably the crowd at the Roma-Lazio game. I expect someone has scored.

(Louder roar.)

My husband is going to kill you and then divorce me if he gets to know about this.

Kohl: I thought you said he was a useless, lazy oaf.

(In Manila, Charlie Bananaboat Deo Deo took out his gun and checked the magazine.)

Gloria: Well, that's it, Kohl. I don't want to ever see you again.

She cut the call and went onto the balcony, to try to salvage something from the mess.

"I have a short statement, to make," she said. "I have been abused and betrayed by a Yankee philanderer. He may have ruined his career, but he's not going to ruin mine. I shall not resign my Chairmanship of the World Body. There is much to do, and I intend to do it.

(Standing ovation from the crowd standing in Parmesano Square.)

I want my husband, Charlie Bananaboat Deo Deo, to know that nothing has changed between us, and I'll send his housekeeping money next week."

(More cheers. In Manila, her husband put away his gun and ordered a Chinese take-out on credit.)

President Sammy 'Stinks' Onions took the accusation leveled by Gloria calmly.

He called a press conference.

"Good evening everybody," he said to the World's press. "It's so nice to see you all so eager to snap up the juicy morsels I am about to cast before you. Contrary to the rumours, I shall be appearing here for the next four years, and look forward to having such a lovely audience on many future occasions. As for my wife in Washington, that bit is true: I do have a wife in Washington. At this very moment she is cooking supper for me, our six children and our little dog, Spit. Now what about the girl friend in every other city in the country? I should be so lucky. I have fans, millions of fans, and some of them are ladies. But does that mean they are my girl-friends? My eighty-year old mother in New York loves me and I adore her. I have a seventy-year old aunt in Chicago. I could go on. Let it be known that when I am on the road, I make a point of visiting my elderly relatives, and am to be seen dining with them in expensive restaurants. So much for my girl-friends. Yes, a question? Saul Woodentop of the New York Post."

Saul Woodentop, the fearless investigative reporter, was in a wheelchair, following an assassination attempt.

"Good evening, Mr President," he said. "Excuse me if I don't get up. I would like to take this opportunity to apologise to the Pro-Life Lobby for my article in the Post. But to another matter: Betty Bone, the Hollywood entertainer, has accused you of making suggestive remarks to her. Is that true?"

"Yes," said Stinks, "it is quite true. When I last took in her act, I suggested to Betty that she should take up some other line of work, or settle down and raise a family. I also suggested that she couldn't sing, dance or act and had a voice like a demented corncrake. Any more questions? Spencer Bronski of the London Guardian. "

Bronski, in his usual dirty raincoat and woollen hat, took his pipe from his mouth.

"Is it true that you are drawing up a blacklist of comedians you consider to be Communist, Trotskyist, Leninist, Stalinist, Maoist, Leftist, Socialist, Democrat, Methodist, Presbyterian or followers of any unpleasant -ism? Is it also true that you intend to stop them performing in public?"

"Spencer," said Stinks patiently, "some of my best friends are Democrats. Can I say fairer than that? The blacklist is only for future reference should the War on Terrorism break out again. Thank you all so much. You have been a wonderful audience. I hope to see you all again soon. Good night and remember my slogan, 'Stick with Stinks, the President who knows his onions."

Stinks was in with a popularity rating of ninety per cent; Kohl Rabi was out, having stupidly resigned. Being a compassionate man, Stinks did not put Kohl out to pasture, he found a job for him.

"But I don't know anything about religious affairs," said Kohl.

"You know about affairs," said Stinks. "You're half way there. As Secretary for Religious Affairs, you simply have to assure all the religions and sects that they are right and everybody else has got it wrong. Whether revelation is carved in stone, cast in gold plates, written on papyrus, parchment, paper, cabbage leaves or the wind, it's the truth. Just keep the parties separated and you can't go wrong. You won't have a problem controlling them, because they'll believe any old rubbish, provided you couch it in archaic English. You'll get to wear a purple suit and meet lots of crazy people."

Rabbit, Coco and I were working flat out following all these developments. Suddenly Coco's legs collapsed and he rolled over. Frank rushed him to the sick bay, and operated on a blown circuit.

"You've been working too hard, Coco," he said. "You'll have to take things a little easier. Avoid stress and watching the news."

I was shocked by Coco's collapse. All this time I had been convinced we were immortal, when, in fact, we were not. It was a sobering thought.

Chapter Thirty-Two

Rear Admiral Hugo Untersea and General Orpheus Quagmire were on the USN Puce, a midget submarine. Being short of stature, both men fitted easily into the two-man cylinder, armed with four torpedo tubes. His face illuminated by the dim, red light of the operating console, Untersea ran through the pre-flight checks. The sub, being the latest model, was fitted with Anti-gravs, which allowed it to proceed to a combat zone at wave-level, thus flying under radar, and not getting wet.

"Are the fish armed?" asked Quagmire, demonstrating his familiarity with Navy vernacular.

"They sure are," said Untersea. "Those four babies could blow most of Panama out of the water."

"Panama isn't in the water."

"You know what I mean, Orpheus. Fasten your seatbelt."

The ship rose just above the waves and moved stealthily towards the buoy where the Nemo was moored.

Rabbit, Coco and I were having a game of 3D chess. Orville and Frank were watching us, while Hiroshi was in the galley, making a supper snack. The boys were really miserable now Rick had run off to marry Guinevere. They took turns to make meals, but none of them liked the others' food. Hiroshi was happy: it was sashimi, beef sukiyaki and Japanese noodles for supper. Rabbit was disgusted. Then the collision warning alarm went off. Orville switched on the floodlights and, through the ports, we got a splendid view of the Puce circling us under water. Orville operated the capture gear and snagged the Puce's nose ring. Midget subs are tiny, light craft, which have very little power. Struggle as it might, it was unable to detach itself from the Nemo. Orville got on the undersea phone.

"Whoever you are, stop struggling. Otherwise, I'll drag you out to sea, take you down to the bottom and watch you implode."

The panic stricken voice of Quagmire came over the intercom.

"We surrender. We weren't doing any harm. We were just out for an evening stroll."

"General Orpheus Quagmire, I presume," said Orville. "I'll give you five seconds to tell the truth. Give me a wrong answer, and down you go, and pop goes your weasel sub."

"We were spying on you," said Quagmire. "I'm a General in the US Army and I'm allowed to do anything I like."

"No you're not."

"No, I'm not."

"Fine," said Orville, "I'll let you surface, still tethered. By the way, what were you going to do with those torpedoes?"

"A little fishing?"

"Wrong answer."

"We were going to blow you out of the water."

"That's more like it."

Orville operated the capture arm and brought the Puce to the surface. We went on deck and watched Untersea and Quagmire extricate themselves from the tiny craft.

"General," said Orville. "You are in Panamanian sovereign waters. I am making a citizen's arrest on the grounds that you intended to destroy the Nemo and kill us all."

"I advise you to think carefully," said Untersea. "The sovereignty of Panama is a fiction. We are the colonial masters. Release my sub at once. Otherwise you will be in deep doodoo."

President Sammy 'Stinks' Onions was being briefed on the Puce Incident, as it came to be known. The Secretary for Religious Affairs was there, dispensing benign smiles on the assembled War Cabinet. Professor Bar Chart advised against another military adventure at that point. The Mars War had emptied the treasury and it was going to take some time for little old Japanese ladies to buy enough bonds to pay for another conflict. Everybody ignored Bar Chart, who was a well-known spoilsport. Farah Share was all for a war. There was nothing like it for killing off men. This would allow more women to get better jobs. Harvey Hacker was reading a comic book.

"Any thoughts, Harvey?" said Stinks.

"Pardon?"

"Should we attack Panama, Harvey?"

"That's the most stupid idea I have heard in several days," said Harvey. "I've got a huge Internet game going, and I'm making a fortune from selling virtual characters to the other nerds. Knock out the three hundred players in Panama and the whole house of cards collapses. Don't do it."

"I have a suggestion to make," said Stinks. "Before we neutron bomb the place and kill lots of innocent doggies like my Spit, who Quagmire has dognapped, I have decided to go to Panama and sort out the problem. After all, 'Jaw, jaw is better than war, war,' as my mother used to say, after giving my father a left hook."

The White House ground crew leapt into action. Water and sewage pipes were disconnected.

Power went over to generators and the ground connections were cut. Air gunners, a recent addition to White House security, climbed into their turrets. The Belly Gunner and Tail End Charlie complained to each other about their lonely, exposed positions. The building rose in the air with a slight creaking and set course for Panama. On the way, the War Cabinet of Speechwriters set about the task Stinks gave them before retiring to take his afternoon nap. They had to find twenty jokes about General Quagmire before nine that evening.

The Interns took turns to fly the White House. Rose Garden, the Intern whose turn it was, made a somewhat wobbly approach and landed heavily in the grounds of the Panamanian Presidential Palace. Stinks was alarmed and asked for a damage report. The belly gunner had been given a fright but was all right. One of the Anti-gravs, however, was badly damaged and would need replacing. There also appeared to be a certain amount of structural damage.

D'Arcy came out to meet the President, and having established that only the White House was the worse for wear, he showed Stinks the Midget Sub, Puce, which was displayed on the front lawn.

"I've never seen anything like that," said Stinks, "It looks harmless enough to me."

"General Quagmire and Rear-Admiral Untersea are with us under temporary house arrest," said D'Arcy. "They'll tell you what happened. Here they come."

"I advise you to declare war on Panama, Mr President," said Quagmire. "Teach them not to be uppity."

"Later," said Stinks. "Show me inside that drainpipe."

Untersea helped Stinks into the sub, joined him in the other seat, and explained the console.

"What are these four red buttons?" said Stinks.

"They fire the torpedoes," said Untersea. "I don't think they're armed."

The President of the United States played the chord of D Molish on the buttons. Four sleek torpedoes leapt from their tubes, slid across the lawn and reduced the White House to matchwood.

"Oops!" said Stinks. "Too bad about that. But it will save on repairs and I can have a nice modern condo built in its place."

Watching all this, Rabbit was on his back kicking his legs, helpless with laughter. Coco, now restored to full health flipped his mouth and eye flaps up and down. I immediately got onto the wire services with the news that the White House had been torpedoed by President Onions. The former location of the White House was invaded by various groups of demonstrators who refused to move until their demands were satisfied. Housefathers were demanding a day a week off; American Indians wanted their country back; Blacks wanted to stop white settlers from taking huge tracts of land in Africa. The US was slipping into chaos.

Over dinner, Stinks, D'Arcy, Quagmire and Untersea, discussed and then signed a non-aggression pact. D'Arcy agreed to accompany Onions back to Washington to do a series of joint charity gigs in aid of the White House Condo Building Fund. The next day, they all visited the Nemo and Untersea apologized to

Orville for the misunderstanding.

"We were only kidding, you know," he said.

"And I was only kidding when I said I'd drag you to the depths to implode," said Orville.

The two Presidents and their families flew to Washington in a 16th century Spanish Cathedral, which was then used as a temporary residence until the White House Condo was built. Stinks and D'Arcy were a huge hit at their gigs, which they put on in baseball stadiums. These had been deserted since all the teams had been sent to China, in exchange for Pandas. The cuddly black and white bear was no longer rare since it had gone on the special TMSG diet. There were now so many, that zoos were turning them down as free gifts, and they were commoner than doggies as household pets. Baseball fans started to ask for their teams and franchises back, but the Chinese refused to budge. The game had become huge in China, with half a billion fans crowding into new stadiums that each held half a million spectators.

D'Arcy and Conchita thrilled their audiences with their exotic Latin American dance. Stinks was at his best hitting the audience with one-liners dreamed up by his speechwriters. His main targets were Kohl Rabi, Gloria Inexcelsis Deo, Professor Bar Chart and the state of the economy. His most hilarious barbs were reserved for General Quagmire and Rear-Admiral Untersea. Big mistake.

Coco got warning of the coup when he intercepted Quagmire's order to his tanks to surround the Presidential Palace and arrest D'Arcy's supporters. Untersea instructed his sub to attack and sink the Nemo, the MV Couch, and any other ships in the port. I immediately informed Brother Heinrich and all ships of the perilous situation. The MV Couch and the Nemo put out to sea, adopted convoy formation with a fleet of Cruise Liners, and set a course for San Diego. Rabbit, Coco and I pulled out all the stops and informed a breathless world that Quagmire had betrayed his country in an Army revolt.

The General went on TV with President Onions' doggie, Spit, which he had kidnapped. He held a gun to Spit's little head and said, "I'm declaring Panama independent and renaming it The Republic of Quagmania. Invade and try to overthrow my government and the dog gets it." Hearing this, Spit bit his hand and Untersea was accidentally wounded when the gun went off. D'Arcy's colleagues once again showed their true colours when they flocked to the Presidential Palace to pledge their allegiance to the new ruler. Such is the true nature of pusillanimous artists.

Chapter Thirty-Three

On the Buddhist Temple heading for Bangkok, Brother Gerbil Sneed, went to the commodious bathing area, stripped off his habit and donned a nifty dark business suit. He placed a neat moustache on his upper lip and a small goatee beard on his chin. Dark shades completed the transformation. When the Temple landed in Bangkok, Brother Gerbil Sneed had become Clinton Ferret, businessman with interests in money all over the world. In Bangkok he rented a villa in an up-market suburb and concentrated on lying low. When he heard that Luigi Parmesano had been brained by a chandelier, he gradually emerged from seclusion and tested the shark-infested waters of life. Leonardo Parmesano was in Sicily and didn't seem to be taking an interest in his father's betes noires. He was concentrating on finding a fresh illegal business now that the people smuggling bubble had burst. Clinton Ferret took a risk and contacted his friend, Brother Heinrich Graft.

"Gerbil!" said Heinrich. "It's so nice to hear from you."

"Clinton, please. My name is Clinton Ferret."

"Still keen on rodents, I see."

"Do you think I would be safe if I came to San Diego?"

"No problem, old friend. You can have your old job back in the bank. I'm so busy playing with my grandchildren that I don't have time to bother with the trivia of high finance."

It was a glorious sunny day when Clinton Ferret boarded the MV Couch. In spite of his disguise, his old acquaintances recognised him at once by the way he kept his hand on his wallet. Back in the Counting House, he relaxed and started to enjoy transferring billions, putting them through the laundry and watching them emerge as pristine cash. He resumed his close friendship with Rick and met his beautiful wife, Guinevere. Amazed by the transformation of the repellently ugly Richard Trickie into the handsome Rick Trichet, Ferret checked into the surgical ward and paid for the full works: nose job, eyes, cheekbones, chin and lips; titanium implants to carry new teeth. His bald head was not a problem, either. The inventor of TMSG discovered that, applied to the scalp, it produced a luxuriant growth of hair. It failed the billiard ball test, but who needed hairy billiard balls? He visited his own mother and she didn't recognize him.

Now that Clinton Ferret, alias Gerbil Sneed, was cosmetically transformed and back in the BwF, Heinrich called on him practically every working day to take him to Megabucks for a latte. Heinrich felt himself slipping back into an earlier time, when he and Gerbil had been men of action. He talked fondly of his days as Gabby Bogart, trekking across the Sierra Madre, searching for and finding treasure. He realized that he had mistaken his faithful donkeys for desperados, plotting to take his gold. He felt a pang of remorse when he realized that he had abandoned them when he escaped in the Supplies-R-Us helicopter. His adventure in Amazonia was still fresh in his memory. He still had the cello, which he had played in the Clinic lost in the jungle. He wondered if Sister Chastity, the Nun, was still there bringing pills and potions to the sick. Heinrich hoped she had thrown out the ones he had concocted to experiment on his patients with. He remembered the great days if the Battle of the Diaoyu Islands, when as Admiral Rouse, he had sunk many a battleship of many a nation. His time as President of Panama was replete with memories and he resented the fact that the history books made no mention of his term. Perhaps there had been so many presidents, there wasn't space to accommodate them all. The gunfight at the Presidential Palace, when he had converted Guido Parmesano to pebbledash with a well-placed tank round, was particularly satisfying.

"Come with me, Clinton," he said. "I want to show you something."

"What is it?"

"It's a surprise."

"I don't like surprises."

Heinrich hailed an Anti-grav taxi outside
Megabucks and told the driver to go to a
warehouse in a derelict area of the port. Clinton
looked round nervously as they passed
crumbling docks and groups of homeless people
standing round blazing oil drums. The taxi
stopped in front of a dockside building complete
with slipway. When they got out, Clinton
noticed that Heinrich was carrying a riding crop.
When a smelly wino stopped them with his hand
out, Heinrich raised his riding crop, thought
better about it, and gave the man ten dollars. He
unlocked the door.

"Close your eyes, Clinton," he said. "And don't
open them until I tell you."

He guided him into the building.

"Right," he said. "You can look."

Clinton looked and beheld an ancient Tiger Tank, still bearing its German Army marks. On trestles, there was a submarine with similar markings and the name U69 on its side.

"Amazing!" said Clinton. "They look as though they have just been made."

"No," said Heinrich, "they are original, but they have been completely restored with the help of my old crew. I have made one modification: I have added Anti-gravs." A head of grey hair appeared in the conning tower. "That's Captain Wolfgang Mozart. He's an experienced submariner with twenty-three kills to his name. He also plays a mean violin."

"Heinrich," said Clinton, "what's going on? Why all the war preparations?"

"Because we're going to war," Heinrich's eyes shone and he looked twenty years younger.

"Who with?"

"Have you been following the events in Panama?"

"I have, and I find the atrocities being committed really shocking. People forced to clock on at work."

"And clock off."

"Magabucks' prices up by fifty per cent."

"Wives forced to stay at home and look after children and do housework."

"Old people force fed tofu burgers with vitamins."

"Babies thrown out of incubators."

"No, not that. You're thinking of Kuwait."

"It's got to stop," said Heinrich. "General Quagmire and Rear-Admiral Untersea are going to get their comeuppance."

Frank and Angela were having breakfast with

Hans and Baby Gretel. They were keeping a careful watch on Hans because he was still resentful of the little intruder. Only the day before, he had given her a rusk, taken her into the park, and left her there. Fortunately Gretel was a messy eater and had left a trail of crumbs, which Angela was able to follow. At the moment, Hans was trying to feed fish bones to his dear little sister. No fool, and used to her brother's attempts on her well-being, Gretel delivered a well-aimed handful of porridge into Hans' eye.

"Angela," said Frank. "Have you realized that this is the second weekend that Grandpa hasn't visited us?"

"When I phoned him, he told me he was working on an important project," she replied.

"Is he in the warehouse?"

"I believe he is."

Divide and rule being the order of the day, after breakfast, Frank strapped Hans into his car seat and drove to Grandpa's warehouse. When he arrived, he was surprised to see the U66 had been launched and the Tiger tank was fastened securely to her deck.

"Going somewhere, Father?" said Frank.

"I wasn't going to tell you, Son," said Heinrich, "because I didn't want to worry you: I'm liberating Panama."

Cyril Liberty, at Central Snoopers, was following the situation. He got a call telling him that Heinrich was preparing an Expeditionary Force. He wasn't too surprised by this news; it wouldn't take much to overcome the garrison, which consisted of fifty men, two tanks, one of which was in the shop for maintenance, and a retired McDuff's Delivery Hound. He put out a general instruction: "Let that mad fool, Graft, carry on, and see what happens. If the situation becomes chaotic, we can move in and clean up."

Frank immediately went to see Orville and briefed him.

"We'll take the Nemo to support him," said Orville.

Not being able to face having raw fish twice a week, he went on board the MV Couch and persuaded Rick to rejoin the Nemo as cook for one trip.

D'Arcy Versey was still in Washington, trying to persuade Stinks to invade Panama and restore him to the Presidency. When he heard about Heinrich's plan, he immediately put together a Concert Party. Then with Heinrich, Frank and Orville, the final invasion strategy was hatched. Prior to the invasion fleet setting out, three small people arrived on the dock: they were the Kalahari Expeditionary Force, Dan, Dana and Bart Bushman. Heinrich welcomed them aboard the U66, thinking such small people wouldn't take up much room.

Gladys Spandau put in the final piece of the jigsaw. Her friends at the Miami Over Sixties Country Club, who were sick and tired of waiting to die, jumped at the chance of speeding up the process. Being rich and vulnerable, they were all karate experts and dead shots with every weapon from a revolver to an RPG. The latter is green and chases you down the street before blowing you to bits. Many a would-be burglar had disintegrated as a result of the accurate use of this weapon. Gladys hired a cruise liner, which had seen better days and was going cheap, and boarded her assault group. The ladies were looking forward to kneeing the occupation troops in the groin. The force was poised like a greyhound straining at the leash.

D'Arcy Versey had his Concert Party in rehearsal, and as usual, was in despair. His Top Hat and Tails Ensemble, his hallmark troupe of dancers, was at it again, stumbling on the stairs, dropping canes and knocking off top hats. Astair Versey, Fred for short, was only six, but he put the veterans to shame. He demonstrated the steps and soon had his elders in shape. The Delightful Dim Sums with Rabbit needed only one rehearsal and almost stopped D'Arcy's heart when Rabbit made a tremendous leap doing a triple somersault and landing lightly on top of the final pyramid. The Cancan Dancers were splendid in their new costumes. The Hungarian Gypsy Orchestra, fresh from Carnegie Hall, played faster than ever, completing Franz Liszt's Hungarian Rhapsody in fifteen seconds flat.

"Right," said D'Arcy to the assembled cast. "We're ready, and the plan calls for us to leave tomorrow. For security reasons, I can't tell you any more."

The Cruise Liner, Sunset Years, docked in Panama port and Gladys Spandau went into the city. D'Arcy's advance publicity firm had plastered the place with posters advertising the The Fol-de-Rols Concert Party, who would be performing for one night only at the Panama City Opera House. Gladys made a block booking in the stalls for her one hundred and fifty Country Club Members. Back on the liner, she organized final training sessions of knee in the groin, Zimmer Frame swinging and cane to the stomach fencing. The girls were ready. The Fol-de-Rols travelled in a regular plane. Wilbur and Mary were with the party to keep an eye on the Dim Sums and stop them doing back flips down the aisle. Coco was with Heinrich on the U66 to act as liaison with rest of the Expeditionary Force. I stayed with Frank on the Nemo. We arrived off Panama and maintained radio silence until the last moment.

The night of the concert arrived and the programme went with a swing. Knowing that they were going to see some action, Gladys' girls were in good voice, calling for encores and giving the chorus boys wolf whistles. For once, the Top Hat and Tails Boys didn't stumble on a stair, drop a cane, or knock off a top hat. The ladies loved them. General Quagmire, and Rear-Admiral Untersea, the guests of honour, and the fifty soldiers and sailors from the garrison were all there in the front row. The Cancan Dancers drove them wild. Then the final production number, involving the whole cast, came to a magnificent climax. The audience was surprised by an announcement that a very special guest was about to entertain them. A roar went up when President Sammy 'Stinks' Onions walked on stage.

"Good evening everybody," he said, "Thank you for that nice reception. You all know me, the President of the United States, who is also a stand-up comic. Every other president was a comedian, of course. The only difference between them and me is that I do it deliberately. I'm delighted to see the General and the Rear Admiral have graced us with their presence and have enjoyed the show. But I haven't come on stage to tell jokes." (Cheers from the audience, who were looking forward to their late dinner.)

"I simply want to introduce some folks you all know and love, your President, D'Arcy Versey, his lovely wife Conchita and their talented son, Astair."

The audience roared a greeting, the Kalahari Expeditionary force popped up from under the seats and pointed their arrows at Quagmire. Heinrich roared up to the front of the opera House in his Tiger Tank. In the meantime, the Nemo was towing Untersea's midget submarine out to sea. The Miami Ladies Assault Force surrounded the soldiers and sailors in the front stalls, and threatened them with Zimmer Frames and canes. No knee in the groin was necessary but several shouts of "Hello, sailor!" were heard. Quagmire, Untersea and the rest of the garrison were warned as to their future conduct, led back to the Canal Zone and locked in. The First Family rode to the Presidential Palace in triumph on the Tiger Tank. The coup was complete and bloodless. Cyril Liberty at Central Snoopers was helpless with laughter. His admiration for his President and D'Arcy was boundless. He swore never to plot against Stinks unless he did something he really didn't like.

It was Fiesta Time in Old Panama. The next day, a great procession formed. The First Family led, seated on the Tiger tank. As in a Roman Triumph, captives (actually cardboard cutouts of Quagmire and Untersea) followed in a tumbrel. The Top Hat and Tails Boys danced along the road in perfect formation, followed by the Miami Over Sixties Country Club. The Cancan Dancers frolicked and flounced causing many a wife to clip her husband round the ear. The Hungarian Gypsy Band played the whole of Brahms and Liszt in twenty minutes and started again. The Delightful Dim Sums, led by Rabbit, making huge leaps and bounds, tumbled merrily on their way.

Afterwards, we all assembled at the Presidential Palace for a Gala Dinner. The Miami Ladies had a terrible time getting ready. During training, they had neglected blue rinses, manicures and general grooming, hoping to inspire terror in the enemy with their unkempt appearance. They were reasonably successful in this, although six of them managed to get dates with sailors. Antoine, the ship's hairdresser couldn't cope and had a nervous breakdown. His assistants did their best to restore some of the ladies' coiffures; the rest had to help each other out.

Gladys Spandau was resplendent in a Chanel evening gown, with jewellery by Heinrich Graft of Hollywood, (So that's where the diamonds went.) and shoes and handbag by Leonardo Parmesano of Palermo, Milan, New York, Miami, Paris and London. Unable to find a profitable illegal business, Leonardo had taken up designing fashion accessories and shoes and was making a bomb. The final word was left to me. I stood on an upturned flower vase on the top table and addressed the company.

"I would like to thank you all for being here this evening to honour the two Presidents and their charming wives. (Applause.) Being a GM Radish is quite good fun but like Coco, Rabbit and the absent Crab, who sends her apologies, I have the Doomsday Gene and can never know the pleasures of family life. Our good friends are our family. (Sympathetic murmur.) Coco and I have witnessed the tribulations and triumphs of our creator, Dr Francis Frank. Lost at six months, he was reunited with his father thirty years later. His father suffered the terrible trauma of losing his wife and son, and, as any normal person would, went quite mad. He scrambled to the pinnacle of business. He suffered the burning desert as a gold prospector. He navigated the Amazon in a leaky steamer. He brought relief to suffering Indians and entertained them with his cello. An hour ago, we received the news that, as the Founder and Patron of Shrinks without Frontiers, he has been awarded the Nobel Peace Prize. (Prolonged Applause.) What can I say of the wonderful Spandau family, except that they are my dearest friends? And as for tonight's guests, if I live to a hundred, and I probably will, I shall never meet finer monkeys-without-tails than you, who are assembled here." (Puzzled silence, followed by rapturous applause.)

www.ingramcontent.com/pod-product-compliance
Lightning Source LLC
Chambersburg PA
CBHW051848170526
45168CB00001B/24